Mobile Strategies

Understanding Wireless Business Models, MVNOs and the Growth of Mobile Content

By

Tom Weiss

Issue Date 15th March 2006 ISBN 0954432770
Published by
Futuretext Ltd
36 St George St
Mayfair
London
W1S 2FW, UK
E-mail:info@futuretext.com
www.futuretext.com

Although great care has been taken to ensure the accuracy and completeness of the information contained in this book, neither futuretext Ltd nor any of its authors, contributors, employees or advisors is able to accept any legal liability for any consequential loss or damage, however caused, arising as a result of any actions taken on the basis of the information contained in this book.

Certain statements in this book are forward-looking. Although futuretext believes that the expectations reflected in these forward-looking statements are reasonable, it can give no assurance that these expectations will prove to be correct. futuretext undertakes no obligation or liability due to any action arising from these statements.

ISBN: 0954432770

This book is dedicated to my Father

For Inspiration,
Education,
and support in writing this book.

Contents

List of Figures

Foreword

By Bob Eggington, Launch Director of BBC News Online

If you didn't know Tom Weiss, you could always locate him in a crowded room. He'd be the one festooned with mobile phones: one in each hand, one coming out of his top pocket and a wire leading from his earpiece to another phone lying on the table.

"Hey, look at this," he'd say with his trademark big grin. Then he would proceed to demonstrate the latest cool thing in mobile.

The thing I always liked about Tom was that although enthusiastic, he never got starry-eyed or carried away with innovation. Once he showed me some new features on some new range of mobiles and asked what I thought.

"How can I put this…I think it's complete rubbish," I replied.

"Oh, they're all rubbish – but they've got potential," he said, with a knowing chuckle.

I spent thirty years of my working life at the BBC and for much of that time I was involved in choosing and managing (or at least trying to manage) a team of specialist correspondents. I developed my own test of what it means to be a specialist.

To me, you are a useful specialist if you have these attributes:

- You have spent a long time thinking about your subject.
- You have brought to that process a systematic mind and a sound judgement.
- You have tested your judgements against your peers.

- You stay up-to-date.
- You can convey your knowledge and insights clearly to others.
- Every time you answer a question on your subject, it is quite clear that you have a vast reservoir of background knowledge waiting to be further exploited.
- It also helps if you have wit, a positive attitude and a sense of history.

So, by these criteria, I love listening to Patrick Moore on astronomy, Andrew Marr on politics, John Keegan on military history, Annie Proulx on literature – and Tom Weiss on mobile telephones.

As long as I have known him, he has been obsessed with communications technology, and particularly mobile technology. He always had to have the latest gadget. Sometimes it seemed the more outlandish the technology the better he liked it.

"Tom, this is fiendishly difficult to use," I told him once, about a mobile device that he had bought as soon as it came onto the market.

"Of course! That's why it's interesting," he said triumphantly. "You must understand that if a manufacturer has worked out every aspect of the technology and the interface and solved all the problems so that it works perfectly, then I have no further interest in it.

"I'm only interested in stuff that presents a challenge – something that's worth a little care and a little effort to make it work. By definition, those are the things that point the way to the future."

As well as devices, he was always keen to understand the technical architectures and the business models. He was the most astute judge of whether some new mobile service or product was going to succeed.

So it was no surprise when he moved to work for T-Mobile.

This period of his employment broadened Tom considerably. It gave him a piece of the mobile jigsaw that had been missing from his experience: knowledge of how the networks played the corporate game.

He must have been good at it, because he rose to become a Vice President in charge of product development. And it gave him unlimited access to new handsets...

But most of all it rounded out his already considerable knowledge about the mobile phone business.

Over the past few years, whenever people have asked me for advice about the mobile sector, I have given them all the same answer: "Ask Tom Weiss." From now on, I shall be able to say: "Read Tom's book."

As I read through it, I was struck powerfully by its depth. I've been talking to Tom about this subject for a long time, but this book revealed much more of the huge store of knowledge he's accumulated.

He is the real thing. If you want to understand mobile, read on.

Introduction

In writing this book, I have attempted to show the opportunities that are available to businesses outside the mobile industry to profit from the mobile revolution.

Primarily for the non-telecoms specialist, it necessarily stays at a high level, focusing on the broad opportunities available to businesses wishing to launch mobile products or services. In order to provide the necessary context of the industry, chapters 2-4 cover the history of telecommunications up to the emergence of early mobile phones. The next three chapters cover the state of the art as we currently understand them, and the final three chapters are slightly more speculative regarding how the future of mobile telecommunications may evolve. I should acknowledge thanks to the following organisations for giving permission to use their material in this book:

- The Marconi Corporation for permission to use photographs from the Marconi Collection.
- The Mobile Data Association (MDA) for the providing the data presented in Figures 44 and 62, and the London Taxi Point case study.
- Esendex for allowing the use of the Parkers SMS Price Check case study and End2End for permitting the use of their case study 'Addressing the mobile-gaming opportunity.'
- T-REGS b.v.b.a. for the French case study on SMS cost regulation, Google for the one on Google SMS and 3G Forum for the Vodafone 3G customers 'Find and Seek' case study.
- Verizon Wireless for the case study on Warner Brothers and Dick Fleming Communications for the case study on getting broadband to the Scottish highlands.

- 12snap for providing an interesting case study on McDonald's and MindMatics for supplying a similar analysis of one of Calvin Klein's marketing campaigns.

- Ofcom, Ovum Nokia and Vodafone for information obtained and to NASA for the illustration of one of their satellites.

I would like to particularly thank Bradley de Souza, Bob Eggington, Hamid Akhavan, and Mehmet Unsoy for their input to the book at early stages in its development, and to my wife Jane for her support in writing the book.

The usual disclaimers apply, in that I am responsible for any errors, and my reviewers responsible for any corrections.

Finally, I have in general tried to use internationally understood English as the standard, with British spelling. The one term for which there is no agreed international standard is the British term 'mobile phone.' The Americans use the phrase 'cell phone' I have used the British term throughout.

Tom Weiss,
London,
January 2006

Chapter 1
Why Mobile?

"Think like a wise man but communicate in the language of the people."
– W.B. Yeats

Since the birth of the telephone less than one hundred and fifty years ago, the world has been dramatically changed by global communications. Over the last ten years, the mobile phone has had a similar level of impact on our lives and our businesses. We are moving into an always-connected world. Information and communication are at our fingertips wherever we are and whenever we choose.

People across the world are taking more calls and most of them are now made using mobile phones. Businessmen dial in to conference calls in their head office from the swimming pools of their hotel on holiday. Transatlantic yachtsmen use satellite dial-up to pick up weather forecasts and e-mail on their laptops, and farmers in Uganda use mobile phones to compare crop prices and prevent their exploitation by middlemen.

This is just the beginning when it comes to the full promise of mobile. Micropayments on pre-paid phone cards are now being trialled in Kenya. In a country where most people are so poor that they do not have a bank account – but many have a mobile – this has the potential to revolutionise the way the business is done. In China, there are no longer plans for the full roll-out of fixed line telephones or Internet to many communities; it is cheaper and more reliable to use wireless.

In the West, the signs of change are even more pronounced. Executives, who only fifteen years ago were primarily communicating by fax, are now hooked to their BlackBerrys, sending e-mails twenty four hours a day. When

I was running T-Mobile's EURO 2004 programme, I remember waking in the middle of night, unable to sleep because of a troubling issue that was not resolved. Rather than roll over and forget about it, it's so much easier to send an e-mail.

The same is true for casual social interaction. When did you last hear people arranging where they would meet? Previously we had to plan exact times and places in order to avoid missing friends. Now we just need an approximate time and location and their mobile number to ensure we can track them down. "Where are you?" must be one of the most common questions over mobile phones: an unthinkable question when using a fixed line.

The next generation of mobile phones is coming with satellite-positioning systems built in. Perhaps the phone itself will be able to answer the "Where are you?" question, and it could certainly give you directions on how to meet up if you both get lost.

In the connected world, a businessman is out jogging early in the morning through New York's Central Park having flown from London the day before, a friend of his calls to arrange a game of squash at the weekend. He takes the call without the friend knowing that he is in New York, and then uses his phone to check availability of the squash court at his local health-club's website. He and his friend agree a time and he books the court. As he carries on running, an alarm goes off on the phone to remind him that tomorrow is his wedding anniversary. He texts the online flower store to ensure that a bunch of roses is delivered tomorrow morning and he checks for a good restaurant to take his wife out to when he arrives back home. In the distraction, he has got himself lost in Central Park, but of course, the phone can direct him back to his hotel.

Is this the stuff of geeky science fiction dreams, or will it become a widespread reality? We have certainly come a long way since the early days of telegraphy in the 1850s. As each generation of technology has been succeeded by the next, each has driven major changes in the way we understand and interact with the world around us. However, the real barriers towards realising these dreams are not the technological requirements, but the appropriate business strategies to realise them.

Ten years ago, no one would have believed that people would communicate with one hundred and sixty character messages in preference to the wonders of pagers or high quality voice calls. Nevertheless, this is now the reality for many of today's youth: and it's big business for those involved in it.

These changes provide opportunities not only for those within the mobile industry, but also for more established businesses to build new product lines or start new services. Virgin, the UK record business, is now the world's most successful mobile virtual network operator (MVNO) and Richard Branson is looking to net more than £500 million on the sale of the business.

Virgin is not the only business that is set to benefit from the increasing mobility of the population. Over the last two years, Apple has established itself as a major player in the music industry from a standing start. Their iPod can only download music when connected to a computer, and there is nothing to say that Apple will also be dominant in online connected music players. Indeed, their sudden entry into this market suggests that it would be equally possible for another usurper to take their place.

Google and Yahoo currently dominate the Internet, but the biggest mobile portals are the likes of Jamba, and Bongiorno. Who will be the winners in the mobile Internet and how will this affect your business? Should publishers rush towards mobile technology in the same way that the Internet inspired mass-investment, or should they steer clear for fear of another dot.com crash?

The differences between mobile and the Internet are significant: mobile is already highly profitable for most businesses involved in it. Many business models for chargeable content are already in place, and there are signs that users will continue to pay to use their mobile phones: something that was never the case in the Internet.

Many successful approaches are already apparent. Companies should now be formulating a mobile strategy to understand how the increasing usage of mobile phones will affect their business. What are the opportunities for increased revenues? What are the threats to their current products or services?

Predicting the future is, as always, notoriously difficult and for every innovation that has taken us by surprise, there have been an equal number that have never achieved the potential that was promised of them at the time. However, when we look at the history of telecommunications, we can see that business models change much more slowly than technology. Indeed, most of the 'new' business models in the mobile industry have been proven over many years in traditional telecommunications business.

To understand future benefits, we first need to understand the past and how the business of telecommunications has developed to the stage that we are at today.

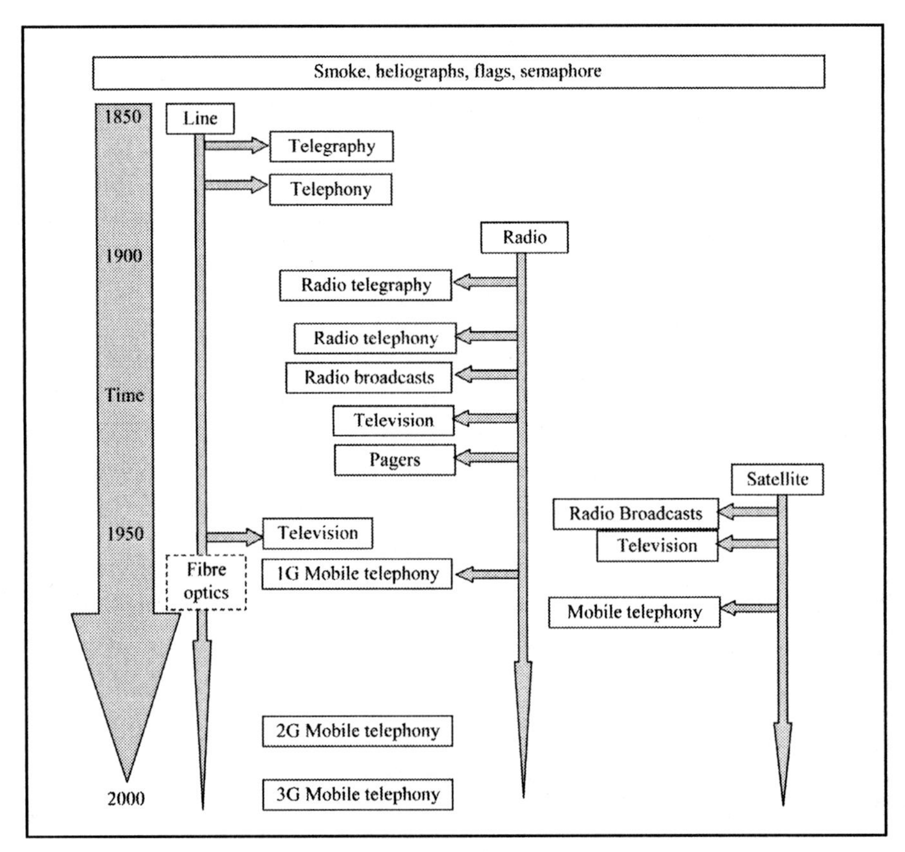

Figure 1. *The development of line, radio, telegraphic and the telephonic communications.*

Chapter 2
The origin of the species

"Until recently the great majority believed that species were immutable productions, and had been separately created. Some, on the other hand, have believed that species undergo modification, and that the existing forms of life are the descendants by true generation of pre-existing forms."
– Charles Darwin

Introduction

The original high-speed data network was a chain of beacon fires used to signal the return of Agamemnon's fleet to Mycenae at the conclusion of the Trojan War over three thousand years ago. In 490BC, at the end of the battle of Marathon, the Greek soldier Pheidippides ran twenty six miles to Athens to announce the victory and then dropped dead from exhaustion. Ever since, improving the speed of communication has been an important challenge. Nevertheless, it was not until the eighteenth and nineteenth centuries that the industrial revolution made for rapid progress and the emergence of telecommunications, as we know it.

Figure 2. *A late Victorian illustration of the signal station at Newhaven.*

Semaphore was an early mobile communications system, initially designed for use between ships at sea. Using the relative positions of two hand-held flags or moveable pointers to indicate different letters and numbers, communication could be undertaken over a great distance. This would only be possible if the two parties were in the same line of sight, with good visibility and, of course, daylight.

Further extensions of these 'visual' systems included heliographs to 'flash' signals if the sun was shining and the smoke signals of the native American Indians. The most famous visual communication was probably Nelson's message to the fleet at Trafalgar, using different coloured flags:

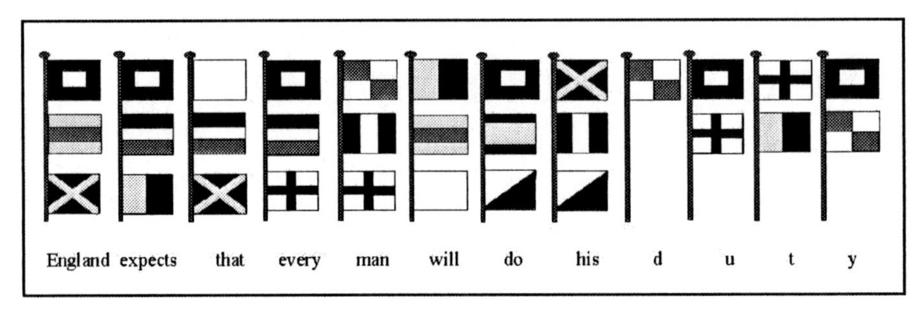

Figure 3. *Nelson's famous flag signal prior to the battle of Trafalgar.*

Telegraph

It was not until 1837, that Samuel Morse patented the electric telegraph, and modern communications were born. Born in Charlestown, South Carolina, Morse was only nine years old when Volta invented the electric battery, and during a period of five years from 1832, he successfully developed his idea that messages could be transmitted by electrical means. More importantly, he recognised that messages would need to be encoded in a form suitable for transmission and subsequently developed the coding system for which he is so well remembered.

Morse's telegraph consisted of a Morse key at one end of the line and a pen attached to an electro-magnet. The electro-magnet attracted the pen to mark a moving strip of paper with a long 'dash' or a short 'dot' at the other end of the line. It was digital system before the term was known. By 1835 he had produced a workable electric telegraph and in 1843 he received a US government contract to construct an experimental line between Washington DC and Baltimore. For the first time, messages could instantly be transmitted by day or night, regardless of weather. By 1861 the east and west coasts of the US had been linked by telegraph and extensive networks were established during the civil war by the armies of both sides.

Throughout the nineteenth century, people continued to look for innovations on the telegraph to make it more user-friendly. Initially the telegram became widespread, automatically printing the telegraphic messages onto strips of text that could be glued to sheets of paper. The Telex, a natural combination of the telegraph and the typewriter, was a far better solution, and first appeared during the First World War, continuing to be widespread until the 1980s and the birth of the fax machine.

As well as improving the technology, people desired to communicate over ever increasing distances, and the applications for international communications soon became clear. In 1865, the first International Telegraph Convention was ratified by twenty participating countries and in the following year the first telegraph cables were laid across the Atlantic Ocean. As soon as it was complete, a brief message was sent simply saying 'All right.'

A few days later the US President sent a four hundred and five-letter message to Queen Victoria; transmission taking just eleven seconds. This performance was considered astonishing. The telegraph continued to deliver telegrams in the UK until these were finally abandoned by BT in 1981, but long-distance submarine cable continues as a practical proposition today.

Telephone

Ten years later, in 1876, Alexander Graham Bell patented the telephone. A Scotsman living in the US, he was the first to transmit speech electrically from one point to another. It is almost impossible not to wonder what Bell thought initially about his invention. "Let me telephone someone," he might have thought. "Whoops, no one else has a telephone." There is a similarity today for those who are irritated by elderly relations who do not have a mobile phone or an e-mail address. His success was helped by Edison's production of a practical microphone and Siemens' invention of the loudspeaker.

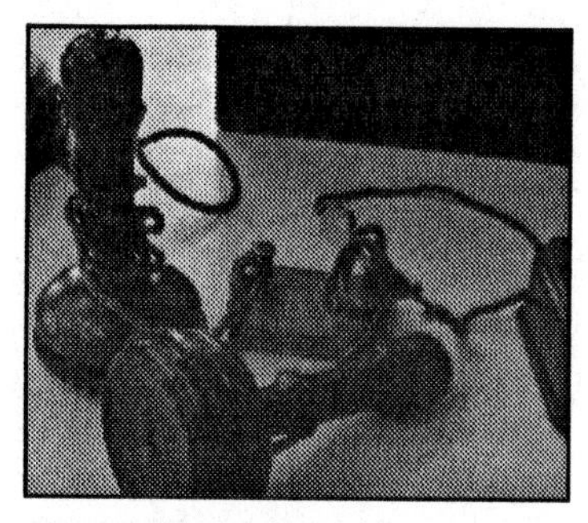

Figure 4. *An early Bell telephone.*

The telephone grew at a phenomenal rate, even taking into account two world-wars during the first half of the twentieth century. However, it is was not until 1946 that a telephone link between the UK and mainland Europe was established and a further ten years – ninety years after the original telegraph cable – that the first transatlantic telephone lines were laid.

The initial transatlantic cables provided thirty six simultaneous telephone channels, and just two decades later, cable capacity had increased a further hundred fold. Over a forty year period, the number of transatlantic telephone cables rose to almost thirty, running underwater from Scotland to Newfoundland, then overland to Terenceville and finally underwater to Nova Scotia.

However, the technical difficulty was not only in laying such large cables. As the telephone signal passed along the wire, the quality degraded to a point that it was no longer functional. The solution was to develop 'repeaters' which amplified the signal every hundred miles or so. The problems in developing adequate repeaters that could operate in deep, salty water, and stay reliable over an economic period – typically twenty plus years – were large.

In addition, the technical hurdles during the early days of telegraphy and telephony were very labour intensive. For the telegraph, human beings had to transmit data in Morse code and regular repeater stations involved operators retransmitting messages, which were of course eventually decoded by another human operator. Telephone exchanges were filled with operators, usually female, who asked the caller the number required, established a circuit and 'rang the bell' of the person being called; a lengthy procedure. Early automatic telephone exchanges were little better. They relied on electro-mechanical operation and were frequently unreliable. It wasn't

Figure 5. *The London General Post Office West telegraph instrument gallery measured some ninety by twenty five metres and housed five hundred instruments, manned day and night by male and female operators though only males were required to work at night.*

until the introduction of subscriber trunk (toll) and international dialling with electronic exchanges in the 1970s and 80s that the situation improved.

Nevertheless, these telegraphs and telephones had three further and serious deficiencies. The first was that the lines were expensive to lay and prone to damage. The second was that they could not be used on moving platforms and this was a serious problem for those who sailed the world's oceans, flew aircraft or drove in vehicles.

The third, and probably the most damaging, was that most industrialised nations had only one telephone company; AT&T in the US, BT in Britain and NTT (formed post-World War II) in Japan. Without competition, these companies all provided the handsets, exchanges and lines, frequently with poor levels of customer service and little regard to the cost charged to the end customer.

The start of competition

The monopoly of these companies was finally broken in the early 1980s. BT was privatised in 1981, AT&T in the US was deregulated in 1984 and Japanese NTT in 1985. Following deregulation, commercial companies were free to establish competing products and services with the incumbent suppliers.

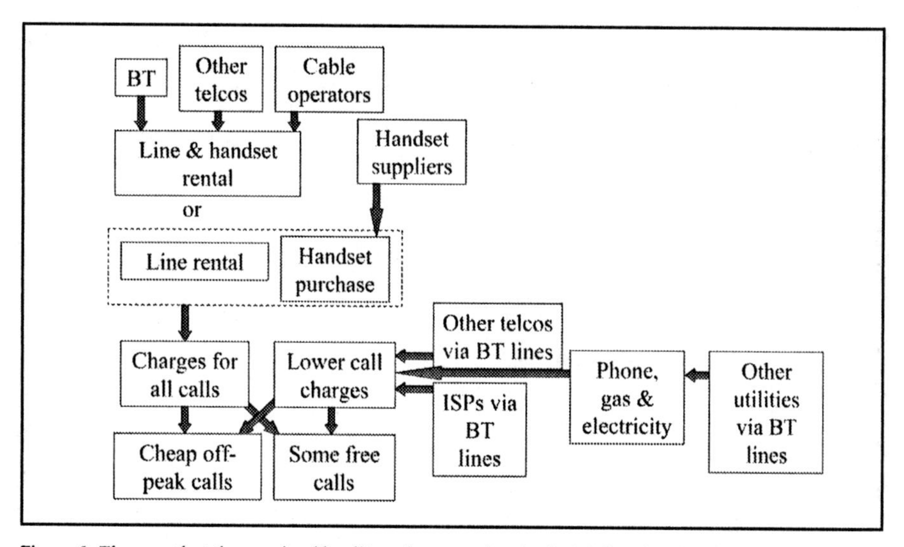

Figure 6. *The way that the supply of landline-phone services in the UK has developed.*

Earlier business models, under monopoly conditions, had involved a rental charge for the line and handset, with call charges based on distance, with reduced calls within a local area of typically thirty kilometres radius. In the US, local calls were free and charges were made only for toll calls. International calls were operator connected and were expensive to make. As electronic telephones were introduced and international calls could be dialled by the user, the cost of such calls fell. Additional services provided included directory enquiries and provision of an accurate speaking clock. However, it was the deregulation of telecommunications and the subsequent competition, that lead to a dramatic reduction in the cost of making calls and the introduction of a range of different business models and services.

UK commercial telephone companies

Mercury was one of the first commercial companies to challenge BT after its monopoly in the UK was lifted. Offering indirect access to the telephone network by dialling a three digit code before dialling the phone number, they were able to offer cheaper calls using existing BT lines. Their 131 service launched in 1986 providing savings on long-distance and international calls. Within a decade, their network was accessible to 90% of the UK population and Mercury customers were offered a phone with a blue button to access the Mercury Network before dialling the number. Further innovations were the option for customers to divide their telephone bills into different categories based on different users or uses. These unique identification codes proved particularly successful where groups of people shared an office or residence.

Another competitor was Onetel, who launched in the UK in 1998, and has grown its provision of both fixed-line and mobile phones to become popular with home users. In 2000, it set up one of the first web based products for customers to manage their accounts online; a complete service allowing anything from paying a bill to troubleshooting. In the summer of 2001, Centrica, one of the UK's largest public utilities, purchased Onetel and in December 2005, when they sold the business, the acquisition provided one million active customers with indirect telecoms services.

The telecommunications business is for the most part capital, labour, asset, and process intensive. The older carriers tend to be vertically integrated with little outsourcing and are likely to operate traditional business models because of their regulatory past. The newer companies are not encumbered by the past requirements of running a vast network. Subsequently they are able to package together highly attractive product sets with focused pricing that frequently undercuts those offered by the BTs of this world.

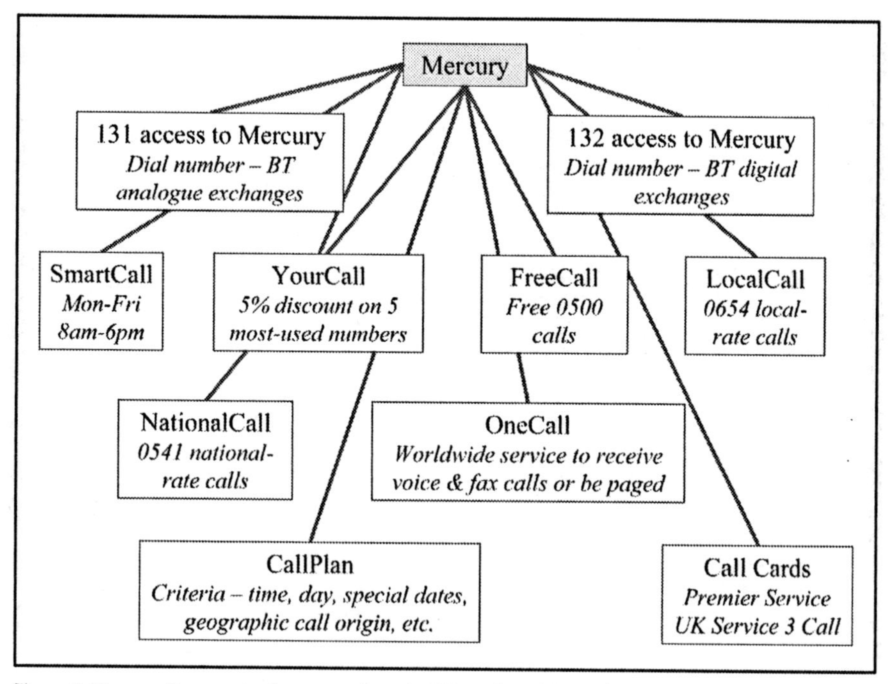

Figure 7. *Mercury Communications opened up the UK market with a wide range of customer choices.*

Case Study: Mercury's pricing model

Mercury's SmartCall offered users a single charge rate for all UK long distance calls using just two time bands; Monday to Friday 8am-6pm, and any other time. Users of SmartCall could also join YourCall Plus giving a 5% discount on the most used five numbers. Premium Call was a service for businesses providing premium-priced information such as stock exchange prices or entertainment. The revenue generated was divided between Mercury and the company supplying the information. Mercury's intelligent network made it possible to route calls to different destinations according to one of seven Customer CallPlan criteria. These depended on time, day, special dates, geographic call origin and other similar parameters.

FreeCall (0500) and LocalCall (0645) numbers were set up to allow customers to call a business either free or at the local rate, with the business paying the remainder of the cost. NationalCall allowed people to call a business at a national rate on an 0541 number.

A Mercury Calling Card was a phone charge-card that enabled the holder to make domestic or international calls from any phone connected to the public network in a participating country. There were three Calling Card services; Premier Service, UK Service and 3 Call. The last proved popular with parents of students who wanted to allow their children to phone home without running up huge phone bills.

Mercury OneCall was aimed at the mobile professional and offered an integrated service. Each user had a single OneCall number on which they could receive voice and fax calls or be paged, regardless of their location, as calls could be forwarded anywhere in the world.

One of the more innovative and important services for the subsequent mobile revolution was the development of Interactive Voice Response (IVR) services. Using IVR, a customer uses a touch-tone phone to interact with a database to acquire or enter information. The technology is widely used by banks, credit-card companies, airlines and cinemas to provide customers with up-to-date details without having to contact an operator. Responses to questions are made by pushing the appropriate number on the phone keypad or by speaking easy words like 'yes' or 'no.'

Premium rate numbers were another side-effect of deregulation. Users are charged a premium, typically 50p per minute, to call these special telephone numbers, and a significant share of this money is paid to the receiver of the call rather than the telco carrying the call. The first premium rate numbers in the UK were prefixed 0898 and this number quickly became associated with adult services because of the high proportion of the numbers taken up by businesses offering erotic chat to their customers.

When combined with a premium rate phone number, IVR allows customers to access premium content over their telephone, such as sports results, chat, and adult services. IVR and Premium Rate are the precursors to the latest generation of SMS services and many of the business models which are now common in mobile were established during these early days of deregulation.

Conclusions

By the end of the 1980s, most of the business models that we are now familiar with in mobile were established. However, one critical factor remained, telephones were still not personal and until the invention of the 'wireless phone' there was no hint of the real revolution that was about to take place.

1. Long range calls using electrical signals were first developed in the 1850s and became widespread during the twentieth century.
2. After the deregulation of the telecommunications industry in the 1980s, the range of products and services has blossomed; together with increasingly low prices for customers.
3. Premium rate numbers and IVR became significant businesses for many firms and set the seed for the explosion in SMS services in the early twenty-first century.

Chapter 3
Look no wires

"After a few preliminary experiments I became convinced that if these waves or similar waves could be reliably transmitted and received, a new system of communication would become available possessing enormous advantages over flashlights and optical methods." – **Guglielmo Marconi**

The discovery of radio waves by William Hertz in 1888 opened the door to the possibility of wireless communications. It took Guglielmo Marconi only seven years to realise the potential of this new technology in its first practical application: the radio-telegraph.

Figure 8. *Marconi inside the station at St. John's, Signal Hill, Newfoundland, after receiving the first transatlantic wireless signal in 1901. (Courtesy the Marconi Corporation)*

In 1895, he demonstrated wireless communication over a distance of more than a kilometre. The next year, in England, he formed the world's first radio company that became the Marconi's Wireless Telegraph Company Ltd. He obtained a patent for telegraphic communication that allowed Queen Victoria to send a message to the Prince of Wales aboard the Royal Yacht.

Marconi's equipment was primitive but using an aerial and an earth connection enabled him to increase the transmission distance to the extent that in 1898 he sent signals across the English Channel. By the end of the nineteenth century, radio-telegraphy was in use both by the Royal Navy and the press. In 1901, transatlantic communication was established from Cornwall to Newfoundland, and in 1918 with Australia.

By 1905, international wireless transmission was widespread. Two early radio successes hit the headlines. The arrest of the infamous murderer Dr Crippen and his mistress was a result of a wireless message from SS Montrose to Scotland Yard. Furthermore, when the SS Titanic struck an iceberg and sank, the survivors owed their lives to the ships that responded to its wireless distress calls.

The use of the term wireless was quite natural bearing in mind that it was an alternative to the wire-connected telegraph. It was not until the early twentieth century that the word radio began to replace it. All the procedures used with the telegraph were naturally transferred to the wireless and many of the early wireless operators had learned their skills working the telegraph.

The success of Marconi's business was not only his successful application of technology, but also his ability to provide easy-to-use products for his customers. Radio equipment was, at this time, difficult to operate and incredibly unreliable. The valves it used failed regularly and were far from stable. The operator controls were complex and it was exceedingly difficult to tune the radio in. As well as providing the equipment for both the land

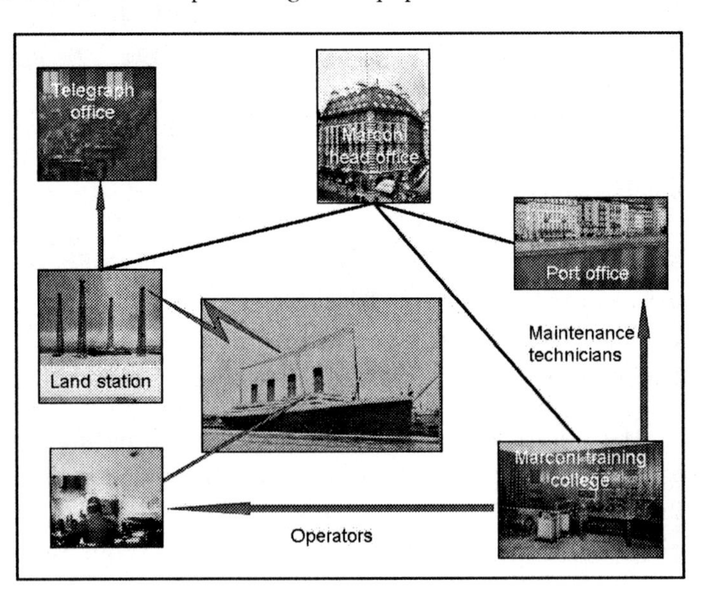

Figure 9. *The Marconi business model was based on the need to deal with the complexities of operating and maintaining early radio equipment. The illustrations show the SS Titanic with its radio aerials and operators cabin. (All pictures except the telegraph office and port courtesy the Marconi Corporation)*

stations and ships, he also provided operators who were capable of running the radio and performing simple repairs. Each major port had a Marconi office with technicians able to carry out the more complex technical repairs that could not be performed by the operators themselves.

We can thank Marconi for the development of practical long-range radio communications between both fixed sites and ships, and later to aircraft and land vehicles. Unfortunately, the equipment was primitive, heavy and temperamental. It was suitable for telegraphy but not for telephony: the bandwidth was fundamentally limited by the very low frequencies that could be generated and it was not until 1914 that products that echo modern mobile devices first begin to appear.

Radio telephony

World War I drove further improvements in communication: development work on higher frequencies was frenetic, as were improvements in the reliability and performance of equipment. In 1914, speech was transmitted by radio for the first time, and by the end of the conflict, a few military pilots could use radio-telephony to talk to their commanders on the ground. However, all conversations were one-way and involved the speakers using the word 'over' to signify that they had stopped speaking and asking for the other party to reply.

Figure 10. *Primitive pre-World War I portable Marconi radio equipment for transportation on horseback.*

The 1920s saw an explosion of mobile communications with radio installed in ships, on vehicles and in aircraft. Later this extended to command and control of aircraft in real time using radar to locate enemy aircraft. During the Battle of Britain, this facility was an essential factor for the success of the Royal Air Force. A further rapid growth of radio communication occurred during the rest of World War II. Since then radio has also become a standard fit in police cars, fire engines, ambulances and taxis.

Personal use of radio communications has also blossomed: as early as 1910, the US Congress defeated a bill to prohibit amateur radio experimentation.

Anyone can become a radio operator providing they can pass an exam to obtain a radio licence. This is still a popular hobby despite the growth of competitive long-range methods of communication. Amateur radio operation, known as 'Ham' uses two-way radio stations in the homes, less often in cars or boats, to talk to people around the world using voice, computers, or Morse code. Some bounce their signals off the upper regions of the atmosphere, so they can talk with people on the other side of the world. Others make use of satellites. Hams can transmit pictures using television and even handle messages during many types of emergency, such as during round-the-world yacht races when a boat gets into distress.

Citizen's band (CB) radio became popular in the 1970s, following the allocation of frequencies in the 27MHz band for operation without a licence. Its use became almost a cult in both the US and the UK, where it was particularly accepted for passing on information about traffic problems and hot spots. Because it did not need a licence, its use became popular with taxi companies and delivery services in the UK. While CB still has a dedicated following, it has become harder to use due to large numbers of undisciplined users and frequency congestion.

Ham and Citizen's band radio lead to widespread adoption of wireless communications. The units could be fitted in moving vehicles, but were still too large to carry on one's person and were certainly not 'mobile' in any modern sense of the word. Furthermore, CB and Ham were still one-way communications methods and could not interact with conventional telephone lines.

Satellite telephony

The first widespread application of wireless communications to conventional telephony came when an alternative to under-the-sea cables was provided by the introduction of satellite relays.

In 1945, Arthur C Clarke invented the concept of a satellite that orbits in sync with the world's rotation, staying stationary over a single point on the earth. These geo-stationary satellites, as they became known, can give wonderful communications coverage. John Pierce of AT&T's Bell Telephone Laboratories first realised that a satellite that could carry one thousand simultaneous telephone calls. He speculated that satellites could generate billions of dollars in revenue by comparing them with the cost and capacity of inter-continental phone cables.

Figure 11. *The Advanced Communications Technology Satellite was launched to provide test communications cover over much of the US. (Courtesy NASA)*

The Russians launched the first satellite, Sputnik I, in 1957. The US responded rapidly. By 1964, four medium-orbit and two synchronous satellites were operating. Following the US legislation in 1962, the CommunicationsSatelliteCorporation (Comsat) was formed and three years later launched 'Early Bird', its first satellite providing global satellite telephone communications as well as a TV service.

In 1964 the International Telecommunications Satellite Organization (Intelsat) was created, which later assumed ownership of the satellites and the task of managing the global system. Much early use of the Comsat/ Intelsat system was for NASA communications. The Intelsat III series, the first to give complete global coverage, was ready just in time to provide near real-time TV cover of the pioneering Apollo moon landing. From humble beginnings – a small number of telephone circuits and a few members, Intelsat grew to provide hundreds of thousands of telephone calls across the world. Consequently, as the number of calls grew, the cost per call fell some hundred-fold.

By the mid-1960s, compared to submarine telephone cables, satellites provided a ten-fold capacity increase for one tenth of the price. Many industrialised nations and not a few developing ones now operate domestic satellites. Each year some dozen or more communications satellites are launched at a cost of some $150 million each. This represents a multi-billion dollar business. Earth station business is of a similar size, as are the communications services the satellites provide.

Radio broadcasting

Broadcasting technology evolved in parallel with radio communications, but with much greater penetration amongst the community at large. The first radio broadcasts in the US started in October 1920. Growth was rapid, with well-known broadcasters such as NBC and CBS starting out during the 1920s, primarily broadcasting local stations across different parts of the country.

In the UK, the Post Office created a number of experimental local stations in 1922, before the BBC, in October that year, and within three years, its broadcasts could be heard across most of the UK. Unlike the US, the BBC was granted a monopoly and was entirely funded by the taxpayers, with no commercial broadcasts permitted.

Broadcasting in Japan started in Tokyo in 1925 and the next year NHK was founded based on the BBC model. A second radio network started five years later with an overseas shortwave service in 1935. All broadcasting was nationalised just before Japan entered World War II and in 1950, NHK became a special public corporation under Japanese Broadcast Law.

The earliest radio receivers were home-made crystal or 'cat's whisker' sets, allowing the listener to hear though headphones. Amplifiers enabled a whole family to listen. Within a decade, many UK homes had mains-powered sets housed in elaborate wood or art deco Bakelite cabinets. On Christmas Day 1932, King George V gave the first royal broadcast to the nations of the British Empire.

The Dutch were the first to broadcast outside their own borders, with shortwave transmissions to the Far East as early as 1927. The Soviet Union followed suit in 1930 and the UK launched the Empire Service, subsequently renamed the BBC World Service, two years later. Today it continues to broadcast in forty three languages to some one hundred and fifty million people across the world. The service is separately funded from other BBC programmes by direct grant-in-aid from the UK government's Foreign and Commonwealth Office.

During the Second World War, the need to communicate to the population became an issue of national security, and listening to the nine o'clock news became a daily ritual to hear about the progress of the war. Radio also became the tool of propaganda, with both the Allies and the Nazis broadcasting into enemy territory, with Josef Goebbels undoubtedly becoming the master of this new medium.

By 1939, the US was the only major nation without a government-sponsored international radio service. The foundling US Foreign Information Service (FIS) was started in 1941 and within months of the entry of the US into World War II, the renamed Voice of America was transmitting to Latin America, Asia and Europe. Today it broadcasts in more than forty languages.

In the post-war era, commercial radio started to have an impact in the UK, with Radio Luxembourg transmitting from the Continent and later, Radio Caroline operating illegally from a ship just outside UK waters. Both were funded by advertising. Today, there is a plethora of commercial stations earning their living from broadcasting interspersed with advertisements.

In the post-war era, relatively heavy battery-powered valve sets were developed, capable of picking up the signals broadcast by overseas stations. However, it was the Japanese who made 'transistor' a household word with their mass production of small and easily portable battery-powered transistor radios.

In the 1950s, because of customers' increasing demand for high fidelity (hi-fi), the use of VHF with frequency modulation and its stereo capability improved the quality of sound broadcasts to a remarkable extent. However, it required a far larger number of transmitters due to the relatively short range of each.

In the early twenty-first century, we are in a position where the number of stations is growing exponentially through both Internet radio and Digital Radio broadcasting. The BBC offers eighteen different digital services across the UK, listen-again over the Internet for the previous seven-day's shows and hundreds of thousands of international radio stations broadcast over the Internet to any country that may choose to listen.

Radio pagers

Radio pagers provide an unusual example of one-way communication to an individual or 'narrowcast', rather than the usual broadcast application of the technologies. However, although they occupy a relatively small niche in today's communications market, they are an important development that provided a solution to many of the early needs for truly mobile communications.

A pager is a dedicated radio-receiving device that allows its user to be sent messages broadcast over a special network of base stations. This is a one-way communication system that allows a single person or a group of people all to be alerted at the same time. Because the pager is only a receiver, it is small, simple and reliable with a very long battery life. Early pagers had no display or message store, but were easily carried and available to warn the user when a message arrived. Today they allow text and voice messages to be sent out and they are very widely used by emergency services and hospitals as well as by local authorities, political parties and some commercial organisations and industrial companies.

The first recorded use of a pager-like system was in 1921 by the Detroit police department. Early paging systems in the mid-1950s in the UK used a magnetic loop around a building and operated at very low frequencies. A UK system was installed in 1956 by Multitone at St. Thomas's Hospital, London. Later the technique changed to VHF radio-based systems but they were only used in on site applications due to their limited range. The term pager did not surface in the US until 1959, when it was applied to a Motorola device. This was followed in 1974 by a pager similar to today's models. By 1980, there were over three million pagers in use worldwide.

A decade later, wide-area paging had arrived and over twenty million pagers were in use. In the next four years, the number tripled and pagers became accepted for personal use. Paging systems have now been developed for monitoring applications and enable the checking of essential information from alarm, monitoring or control systems, such as production data and temperature changes. However, wide-area paging systems have largely been replaced by SMS texting. In North America, paging became so extensive that it actually delayed the widespread use of mobile phones, and Research In Motion, the company that sells the popular BlackBerry product was initially a pager company.

The main advantages of pagers, when compared with mobile phones, are the low cost of delivering messages, the broadcast capability allied to controlled accessibility and reliability, and defined coverage. In addition, the battery life of pagers is far greater than that of mobiles. The major advantage of paging over SMS comes when a message needs to be sent to many recipients, when pagers are faster and less expensive.

Comment: The development of electronics

It can be difficult to understand some of the changes that have occurred in communications without an insight into the technology advances that drove them. The electronics industry was a twentieth century phenomenon that forever changed the way people interact and are entertained, and the way armed forces fight. Telephones and mobile phones, radio, television, radar, computers, satellites and process-control systems are all children of the changes introduced. In the first half of the century, solutions were often unwieldy and difficult to use, but all this was set to change, though many argue that the growth in functionality has resulted in equipment that is still hard to operate.

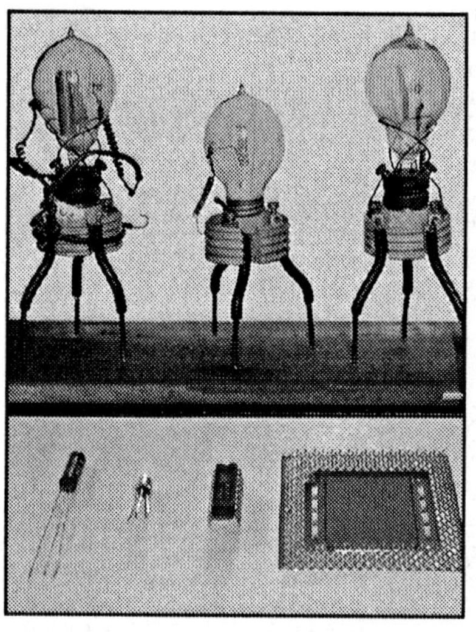

Figure 12. *Fleming's original valves (courtesy the Marconi Corporation) of fragile metal in glass construction, compared with from left to right an early germanium transistor and a silicon transistor that provided equivalent functionality, an integrated circuit that contains tens of transistors and a microprocessor with tens of millions of transistors.*

Sir John Ambrose Fleming is considered by many to be the founder of electronics. While professor of electrical engineering at University College London in 1904 he invented the thermionic valve (or vacuum tube to Americans), so named because, like a water valve, it allowed electronic flow only in one direction. His invention enabled wireless technology to become a practical proposition. However, the valve remained a delicate and power-consuming device for the next half century. The valve was widely used in radios, TVs and early computers until an alternative became available.

The replacement of the valve by the transistor had a fantastic impact on the design of electronic equipment, due to its robustness, long life, modest power consumption and fundamental low cost. The impact of the transistor (the first ones were made from germanium and were electrically fragile and were soon replaced by silicon) was to make much valved equipment obsolete virtually overnight. Compared with the valve, the transistor provides the same function in a tiny durable package that can be powered by a small battery. The technology led to the production of the first practical mobile telephones.

The integrated circuit (IC) was invented by two Americans working for competing electronics companies; Jack Kilby of Texas Instruments and Robert Noyce co-founder of the Fairchild Semiconductor Corporation. In 1959, both parties were granted patents. After some legal wrangles, the two organisations agreed to cross-licence their IC technologies and the applications mushroomed. The first ICs, which only contained a small number of individual transistors and other components in a single chip, were used in computers, ballistic missiles and hand-held calculators. Today's ICs hold many more transistors and were widely used to help reduce the size of mobile phones during the 1980s.

Robert Noyce went on, with Gordon Moore to found Intel. Intel produced the first universal microprocessor, the Intel 4004 with roughly the same computing capability as the original ENIAC computer, which used eighteen thousand valves occupied one hundred cubic metres of space and weighed thirty tons. The single chip measured just 3mm x 4mm, containing over two thousand transistors in circuits that acted as a central processor, memory, input and output controls. The main application of these new microprocessors was as industrial controllers. However, during the next three decades, the microprocessor was developed to include millions of transistors and became the heart of both the PC and the mobile phone.

An application-specific integrated circuit (ASIC) is a microprocessor designed and manufactured to provide specific functions for an electronic device such as a mobile phone. The circuit design is undertaken, in this case, by the phone manufacturer and the ASIC produced by a specialist foundry. Because ASICs are customised to a particular requirement, they deliver excellent performance and functionality. ASICs are employed in a wide range of consumer and industrial electronic equipment.

It should be apparent that it was the improvement in electronic components in the twentieth century that first made radio communications possible and later led to the mobile phone revolution.

TV broadcast

Television was in many ways a logical extension of radio, although, to date it has had less impact on the mobile phone industry. The emergence of mobile TV and paid for content means that television and mobile business models may yet converge.

Figure 13. *The original 1929 low quality thirty-line Baird TV test pattern. (Courtesy MHP, the test card gallery)*

A Russian, an American and an Englishman all claim to have 'invented' television. The first practical solution was provided by the Scot, John Logie Baird. His first working television system was used for the first BBC broadcasts from Alexandra Palace in 1936. The system, involving mechanical scanning at the studio, gave a flickering black and white picture with only thirty lines on the screen.

There were just twenty thousand homes with a television within a fifty six-kilometre range of Alexandra Palace. The first sets cost a fortune; as much as a small car. Transmissions ceased during the war period and reopened in mid-1946, but by the coronation of Queen Elizabeth in 1953, television had taken off with an estimated twenty two million viewers across the UK.

The Beveridge Committee had recommended that the BBC should remain the authority responsible for all broadcasting in the UK. This led to controversy and then competition in the mid 1950s with the launch of commercial television. The BBC, hampered by its reliance on a fixed government licence fee, saw its audience share drop but the impact of competition forced the BBC it to overhaul its programmes. It was not until 1964 that the BBC launched its second channel; BBC2.

Initial BBC broadcasts relied on four hundred and five line transmissions, while the US had better resolution with five hundred and twenty five lines. The UK then moved to six hundred and twenty five lines for colour and still provides better resolution than US broadcasts. The move to colour started in the 1950s. It was totally accepted within two decades. In a period when disposable incomes were much lower and TV sets were ten times more expensive and relatively more unreliable than today's conventional models, renting became a popular business model. Companies like Radio Rentals provided the TV set and the repair staff, as well as updating users' models at regular intervals – typically every three years. In a business model very similar to that of today's mobile operators, it required a significant up-front investment in the TV sets but resulted in a large return on sales.

The US approach to television was commercial from day one. However, because of the size of the country, the stations and their audiences were restricted to local broadcasts only. There were only some seven thousand sets in use by the end of 1945 and even a year later there were only half a dozen TV stations, all of them located in major cities. The 1947 World Baseball Series resulted in the first mass audience for television; nearly four million people watching, mostly in bars. In 1948 ABC and CBS first started broadcasting 'network' shows and by the mid 1950s colour transmissions were widespread, mostly funded by advertising.

However, pay-TV in the form of cable was just around the corner and by 1958, over five hundred cable TV systems were serving nearly half a million US subscribers. In that year, CBS took out a two-page advertisement in 'TV Guide' magazine warning that free TV could not survive alongside pay-TV. By 1961, in search of added profit, ABC stretched the length of its advertising breaks from thirty to forty seconds. Inevitably, the other networks followed suite. A decade later the time had reached one minute.

CNN (Cable News Network) was established in 1980. By the end of the decade, pay-as-you-view had become a norm for cable TV services, reaching about 20% of all cable users. Audiences for the broadcast networks, meanwhile, had sunk to an all-time low of just over half the total.

In Japan, NHK (and NTV) did not start television broadcasts until 1953 and moved from black and white to colour pictures in 1960. Although NHK also introduced commercial broadcasting, today it is funded by a licence fee, with four other major nationwide Japanese commercial networks.

Pay-TV has not been as widespread in either Japan or UK as in the US mainly because of the dominance of the traditional broadcast networks. However, in recent years this has changed with cable companies launching television services across Europe, and most significantly the entrance of Satellite TV.

Satellite TV

Satellite TV allows the broadcast of a large range of channels to a large audience across a wide area.

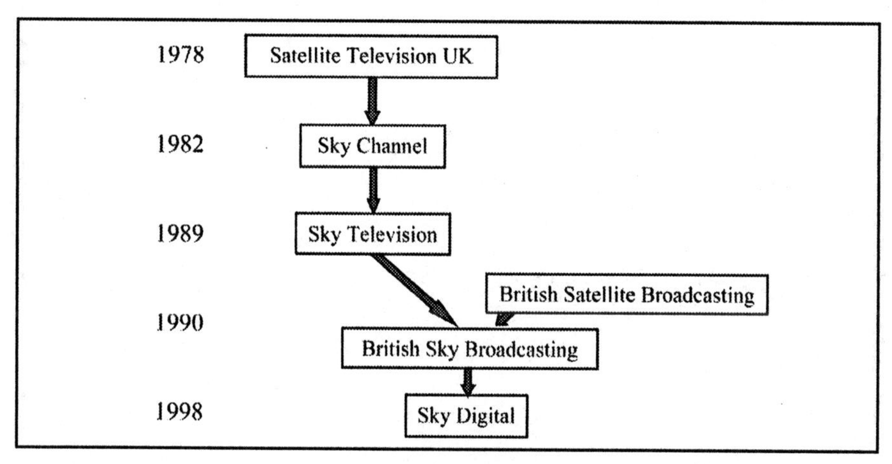

Figure 14. *The development of satellite TV companies in the UK over a twenty-year period.*

Satellite broadcasts started to the UK in 1978 from a relatively unknown business called Satellite Television UK. This was bought by Rupert Murdoch in 1982, and rebranded as Sky. A competing service was launched in 1990, called British Satellite Broadcasting, but it quickly became clear that there was not sufficient market for two satellite broadcasters. The two agreed to merge at the end of 1990, rebranding as BSkyB.

One of the advantages that BSkyB offered was an interactive service. Pressing the coloured buttons enables a range of menu driven options to be selected. At an event like the Wimbledon tennis tournament, this allows viewers to choose which particular match they wish to view from a number of matches being transmitted simultaneously.

Figure 15. *Satellite receiving dishes have appeared on large numbers of homes in the UK.*

The launch of satellites in a new geo-stationary orbital position enabled the company to launch a new service, Sky Digital, the most popular subscription-television service in the UK. In 2005, BSkyB had turnover of just over £4 billion, with a profit after tax of £425 million. It had nearly eight million direct customers a further three and a half million customers subscribing to some of its channels via cable TV.

Sky's business model was based on renting satellites, selling receiving sets to customers and charging a monthly fee for a basic package of channels, with additional channels at extra cost. Today, there are offers of free installation of the customer's equipment (a dish and set-top box) and a half-price fee for the first three months' subscription. In 2004, the pay-TV revenues in the UK exceeded advertising revenues for the first time, and this is mainly due to the success of Sky. Sky's success is driven by three factors:

1. Intelligent pricing – The cost of an entry level Sky system is low, but once acquired it is easy to upgrade to premium channels.

2. Reliable, easy to use technology – The set-top boxes and satellite dishes are reliable and the software installed on them is intuitive to even the most technophopic.

3. Exclusive content – As well as famously buying up the rights to football[1]

[1] It is a quirk of consumer sentiment that more Sky customers have registered 'football' as the primary reason for purchasing Sky than actually watch football on a regular basis.

matches, Sky screens most US dramas at least six months ahead of other broadcasts.

In addition, Sky has been remarkably successful at buying the TV rights to major and therefore popular sporting events.

Satellite television has also launched pay-TV services in Japan. Following the launch of a basic service in 1984, regular broadcasts started in 1989. By the Millennium, Japan had ten million satellite viewers, and digital terrestrial transmissions started in 2003.

Conclusions

The development of wireless communications and broadcasts has many parallels with the modern mobile phone industry:

1. The early Marconi model was based on providing both the equipment and trained operators at the land-based stations and on board ships, with repairs on the latter carried out by staff at Marconi port offices.
2. Equipment unreliability in the consumer field led to the growth of rental companies in the early 1950s, particularly in the TV sector, with a large investment in equipment and a large profit on sales.
3. The introduction of satellites instead of submarine telephone cables provided a large capacity increase and a ten-fold price reduction. This has grown to today's multi-billion dollar business.
4. Satellite TV business models are based on a mixture of subscription fees and advertising revenues, with subscription fees for television now exceeding advertising in most markets.

Chapter 4
The mobile revolution

"The total market for mobile cellular phones will be nine hundred thousand subscribers by the year 2000." – **McKinsey Study for AT&T, circa 1980**

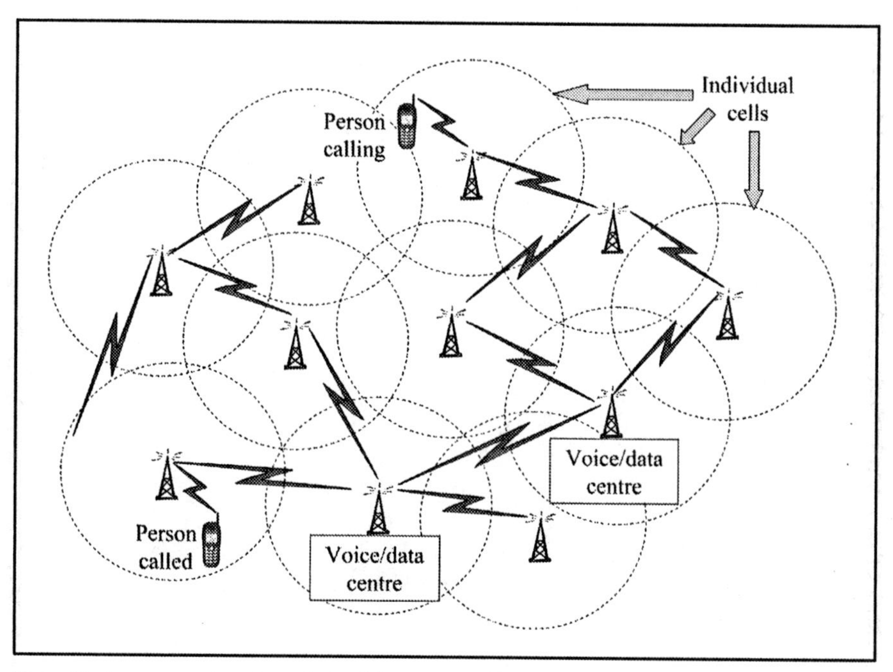

Figure 16. *A representation of a typical mobile network setup.*

Anyone who wanted to communicate on the move before 1975 faced cumbersome equipment that was generally unreliable and not particularly portable. There was clearly a fortune to be made by whomever could overcome these difficulties. As early as 1947, the size of suitable equipment was reduced to the scale whereby it could be installed in cars. That year AT&T asked the US Federal Communications Commission (FCC) to allocate sufficient frequencies to allow a wide-spread mobile phone service to be developed. However, the FCC was only able to allocate enough for around twenty phone conversations simultaneously in a single area. In principle the system was similar to modern services, using many base stations to divide a service area into cells, and transferring calls from one base station to another as users moved from cell to cell.

The first improvement was to increase the number of calls that could be made simultaneously. In the late 1960s, AT&T proposed a new system using many low-power broadcast masts, each covering a radius of a few miles but collectively covering a larger area. Each mast used fewer frequencies and as cars moved across the area their calls passed from mast to mast. The FCC approved this increase and these phones were fitted extensively among the emergency services as well as to taxis.

Figure 17. *Mobile telephony 1924 style. (Courtesy Bell System Memorial)*

It took another decade before the first demonstration of a mobile phone that was small enough to be carried around on one's person. The Motorola device weighed 1kg and was trialled in Chicago with some two thousand customers. In 1981, a second cellular system was tested in the Washington/Baltimore area and the following year the FCC finally authorised a commercial service, which was launched in 1983, with phones weighing 500g and costing $3,500. Despite the huge need, it had taken nearly four decades for mobile phones to become technically feasible and commercially available in the US.

This first service was plagued by unreliability and the limited number of subscribers that could be supported by the available frequencies. After five years in operation there were one million subscribers, which may sound significant, but by the standards of American telecommunications was not even a starting point; it appeared that the business could not make economic sense.

In the meantime, Japan and Europe were pushing ahead. Unlike the US, there had been no widespread deregulation of the fixed-line markets, and mobile telephony offered an opportunity to introduce real competition into the marketplace. Mobile phones appeared in the Tokyo area in 1979 and in the UK in 1985.

The global explosion

The birth of modern mobile telephony occurred in Finland in the late 1970s when Nokia and the TV manufacturer, Salora, combined forces to develop a new mobile phone. This was to stimulate Europe to introduce a new digital wireless standard called GSM. The world's first international cellular mobile phone network (NMT – Nordic Mobile Telephone) opened in Scandinavia in 1981 with Nokia car phones, using analogue technology at UHF frequencies. The company produced its first hand-portable mobile phone, the Nokia Cityman, six years later.

Things really took off in Europe with the launch of GSM in the early 1990s. This is the digital technology that underpins the majority of 2G networks. It allows eight simultaneous calls on the same frequency and its use is widespread across most of the world, employing frequencies in the 900MHz, 1800MHz and 1900MHz bands. There is no question that the introduction of GSM led to an explosion in mobile phone usage.

Meanwhile in Japan, NTT, the state telecommunications company, had a monopoly in mobile phones throughout most of the 1980s. Although this monopoly formally ended in December 1988, it took a further five years for the market to blossom with complete deregulation, a move to digital technology, price breaks, and customers actually owning their own phones.

The technology (PDC) was different in Japan: the radio networks could transmit over much shorter distances than their American and European counterparts. Although this meant that more base stations were needed, the lower requirements on the handset meant that the Japanese phones of the 1990s were much smaller and lighter than in other countries, and had a significantly longer battery life. Combine this with a roll-out of the network across the underground-train systems, a nation of commuters spending long hours away from their small homes and you have the ingredients for the explosion in mobile telephony that first broke in Japan.

So why did different regions of the world end up using different systems? Firstly, the solutions to the problems of mobile telephony were developed independently in the different parts of the world and at a time when global communications were far more restricted than they are today. Furthermore, the geographies of the regions vary considerably, as does the physical size of the industrialised nations involved. Finally, there were the competitive pressures of the companies involved. Qualcomm, for example, support CDMA. They lobbied heavily for the support of WCDMA in the US to ensure that they would continue to have the edge over European manufacturers such as Nokia and Ericsson in the market for US network equipment.

It is unfortunate that this fragmentation of standards has become a common feature across the mobile industry. Compared with the PC industry, where the Wintel[2] coalition caused dramatic standardisation in the early 1990s to be cemented by the web standards of the late 1990s, the mobile industry is a mire of differing network protocols, application standards, programming environments, and development tools.

2. PCs using Microsoft Windows operating systems and Intel microprocessors.

The three examples from the US, Europe, and Japan show how choices of underlying technologies play an important part in the success of mobile operators. Standards have been grouped into a number of generations associated with both data transmission speed and quality. In each of these generations, there have been a number of competing, but incompatible methods for maximising the amount of data that can be transmitted over a network. In the first stages, the Japanese got clear advantages in their low-power devices while the Europeans had success with GSM. The US however, having invested heavily in analogue systems and having been unable to recoup their investment, were reluctant to move into digital networks. Thus, the USA has subsequently lagged behind in the widespread adoption of mobile technology.

Generation	System employed	Geography
1G	NMT	Scandinavia
	AMPS	US
	TACS/ETACS	Europe
	NTT	Japan
2G	GSM	Europe, Asia
	PDC, iDEN	Japan, Korea
	TDMA (DAMPS)	US
	CDMA IS-95	US
	IS-136	Asia
2.5G	GPRS	Global
	CDMA2000	Japan, China, Korea, US
	UTMS	Japan, Europe
3G	WCDMA	Japan, Europe
	EDGE	Asia, Latin America

Figure 18. *Some examples of the early generations of mobile phone systems and where they were deployed.*

We should not overstate the technology impact since the cultural differences also played an important part. America had a well established, competitive, fixed-line business which offered most customers free local calls and competitive rates for long-distance. In Europe, buying a mobile phone was often the first time people had access to 'free' minutes and the introduction of competition is equally relevant to the success of GSM in Europe as the technical standard itself.

Once prices had reduced sufficiently to be affordable for private individuals, growth happened exponentially. It is much easier to spend time on a mobile phone than a static one since 'time on the move' otherwise tends to be dead time. A personal phone can be a godsend to adolescent children who no longer face arguments about the amount of time on the phone, and for parents who can make their children pay directly for their own phone calls.

Adding to the benefits of mobility, mobile phones quickly became feature-rich devices with significant advantages over traditional phones. This was partly driven by the rapid improvements in the size and battery life of early phones, but also because in most countries it became common practise for large network operators to subsidise new handsets in order to win customers.[3]

The young people who had previously been given mobile phones by their parents quickly started to experiment with the features and found plenty of additional uses. SMS is the most famous of these with the unique language and subculture associated with it. Less obvious, but more compelling is the use of the address book in the mobile phone. People losing their phone nowadays have often not only lost the ability to be called by their friends, but have frequently lost the very information they need to contact them.

Of course, mobile telephony was still more expensive than calling from a landline, but clever pricing has created the perception of lower cost. Giving one hundred free minutes per month encourages people to talk to that level, and maybe slip slightly beyond. These offers are not without their dangers. In the UK, One 2 One (now T-Mobile) aimed to attract new customers by offering free evening and weekend calls. These phones were very quickly bought up by call centres to reduce their operating costs, while at the same time providing a massive loading on the network.

Mobile phones in developing nations: Cameroon

In many developing nations where there was only a primitive fixed-line network, mobile networks have developed even more rapidly becoming the preferred choice for phone communications across the board.

Cameroon is a large country (approximately five hundred thousand square kilometres) just to the east of Nigeria on the Gulf of Guinea. It has a flat marshy coastal strip but is dominated by tropical rainforest. It has two large cities; the capital Youandé and the main port, Douala at the tributary of the two key rivers. The Mbang mountains run up much of the north-west border and effectively split the country in two. North of the

3. China and the Netherlands are the most obvious examples of countries without handset subsidies.

mountains there are only three sizeable conurbations. The land becomes flat running up to Lake Chad. The population is some eighteen million, of which one million live in the capital.

Basic telecommunications services were run as a state monopoly by the Ministère des Postes et Télécommunications (MPT). It was also responsible for regulating the telecommunications sector. In 1998, two public enterprises, Camtel (fixed lines, with a waiting time of some six years for a new connection and only one hundred thousand subscribers at the end of 2003) and Camtel Mobile, were set up to take over from MPT and a new regulatory body became operational at the end of 1999. In June 1999, a mobile telephony licence was granted to a private firm. Both enterprises were privatised in 2000 and Camtel Mobile was acquired by South African company, MTN International and renamed MTN Cameroon.

Demand for mobile phones rose dramatically after the liberalisation of the mobile phone sector and the opening of it to competition. France's SCM and MTN Cameroon were the early entrants into the market and launched aggressive campaigns to woo customers. They were quickly joined by Orange Cameroon. It is worth noting that billing proved one of the critical areas due to the difficulties of collecting the revenues. At the end of 2005, MTN Cameroon had increased its customer base to over one million, an annual growth rate of some 25% and claimed to have a share of just over half of the market. It is clear that even in this developing nation, mobile telephones offer a far better solution than fixed lines for the vast majority of users.

This has led to the position in the early part of the twenty-first century where there are more phone conversations each year over mobile networks than over fixed line. The fixed-line telecommunications companies (telcos) in every country have seen significant reduction in their revenues from voice, which for the most part has been mitigated by the growth in fixed-line Internet connections.

No one has to be reminded that a mobile phone still has to be within the radio coverage of their network provider in order to get a service. This can be difficult even within a country, where remote regions can prove a particular problem. With most handsets being available on most networks, during the initial-growth phase competition centred on price plans and coverage. Users would choose their phone depending on how likely it was to work in the vicinity of their home, their work and their friends, and on the price plan that best suited them.

In countries having highly built-up networks like South Korea and the UK, virtually full coverage was achieved by early 2001. However, full coverage will always be a drawback for the larger European countries and the USA, where there is simply no business case for rolling out cellular networks across the Great Plains. But in these areas where the economics of cellular networks do not make sense, several companies committed themselves to providing phone systems using satellites in low earth orbits; at a height of about eight hundred kilometres such as Iridium.

Iridium has almost complete earth coverage using some sixty satellites operating as a single network. The total cost of developing the Iridium system was well in excess of three billion dollars. This is a huge sum – but significantly less than the price of a 3G licence in Europe. Unfortunately the service itself is not as compelling for users since the handsets require much more power to transmit signals to the satellites and thus need much larger batteries.

These services are ideally suited for industries that are forced to operate in remote areas such as maritime, aviation, government/military, emergency/humanitarian services, mining, forestry, oil and gas, heavy equipment, transportation, the media and utilities. In 1994, Vodafone joined a consortium to develop and launch a low-earth-orbiting satellite mobile phone service to enable users with dual GSM and satellite handsets to exploit either the existing land-based network or, if out of cover, the satellite system. There are several smaller organisations offering services that are more limited and this model of multiple networks accessed from a single device is a model that is likely to become increasingly common in the future.

The modern mobile phone

Mobile phones have developed now to the stage where they are not simply a communications device, they are a personal and fashion statement about their owners. This has encouraged the development of new technologies to provide an ever increasing range of functions in handsets. Today's mobile device is a powerful wireless computer that would have astonished PC users only a decade earlier. Standard items include an address book, calculator, calendar, clock with alarm function, and stop watch as well as voicemail and text messages. In the late 1990s, Nokia started selling small, lightweight mobile phones that were viewed as fashionable and appealed to the young. More to the point, they were affordable as well. Just as we now assume that almost everyone has an e-mail address, so we also take it for granted that they have a mobile.

Although a mobile phone seems to be a relatively simple item that is easily carried in a pocket or handbag, it is in fact a relatively complex system in itself that requires a fair amount of miniaturised hardware and software to operate effectively. Yet the majority of users appear to be more interested in the colour and shape of the faceplate than any other factor; they may also be concerned about the size, weight and battery life.

Early phones developed for each of the new mobile phone generations have not been renowned for their reliability, either in hardware or software terms. This has badly impacted on early adopters and has driven some back to more reliable but less capable handsets. As mobiles become more PC-like, there is also a real danger of viruses and the first one, Cabir, was discovered in the middle of 2004. Although there has not yet been any widespread disruption caused by mobile phone viruses, it is inevitable that problems will emerge over the next five years.

While in most countries, the mobile phones are provided by the network operator, the phone manufacturers have still succeeded in building huge brands. The top mobile phone manufacturer worldwide is Nokia of Finland with a market share of almost one third, strong sales in the Far East and one of the world's most-valuable brands. In 2004, the total sales of Nokia mobile phones reached over two hundred million units – that's six phones every second – a growth of 16% compared with the previous year. In its latest financial report, Nokia reports selling over €5 billion worth of mobile phones in the most recent quarter, and the company has suggested that the total market that year was nearly six hundred and fifty million units.

Motorola, the US company, is the second most successful mobile phone supplier with some 15% of the market and with the top brand name in the US and rest of the Americas. Like Nokia, it is seeing strong growth in the developing world and has recently launched an iTunes phone with Apple. Samsung dominates the Korean market. It has some 12.5% of the market and is growing its deliveries in Europe and North America. During early 2005 Samsung managed to push Motorola off the number two spot, but sank back into third place later that year.

The other significant suppliers are:

- Siemens, the German manufacturer recently bought by Taiwan's BenQ.
- LG Electronics with a range of new CDMA and GSM phones that are popular in Korea and the US.
- Sony Ericsson that has been promoting Walkman-phones and concentrating on top-end mobiles. All have significant but small market shares.

There are literally hundreds of different handsets launched each year, mostly with colour screens and cameras. Each phone is carefully targeted at a different market segment: business users want phones to look different from those aimed at younger audiences, and young women will want phones different from those preferred by young men. In addition to the cosmetic differences of mobile phones, there are also a number of different functionality approaches. Each of these is an example of convergence around different consumer electronics devices.

Although these six approaches involve different specialisations, it is worth noting that as phones mature, most of the functions of specialist devices are available on the mainstream handsets. For example, the Nokia 6230, one of the best selling devices of 2005, covers a broad range of features. It has basic Microsoft Outlook functionality (PDA like), includes a FM radio and plays

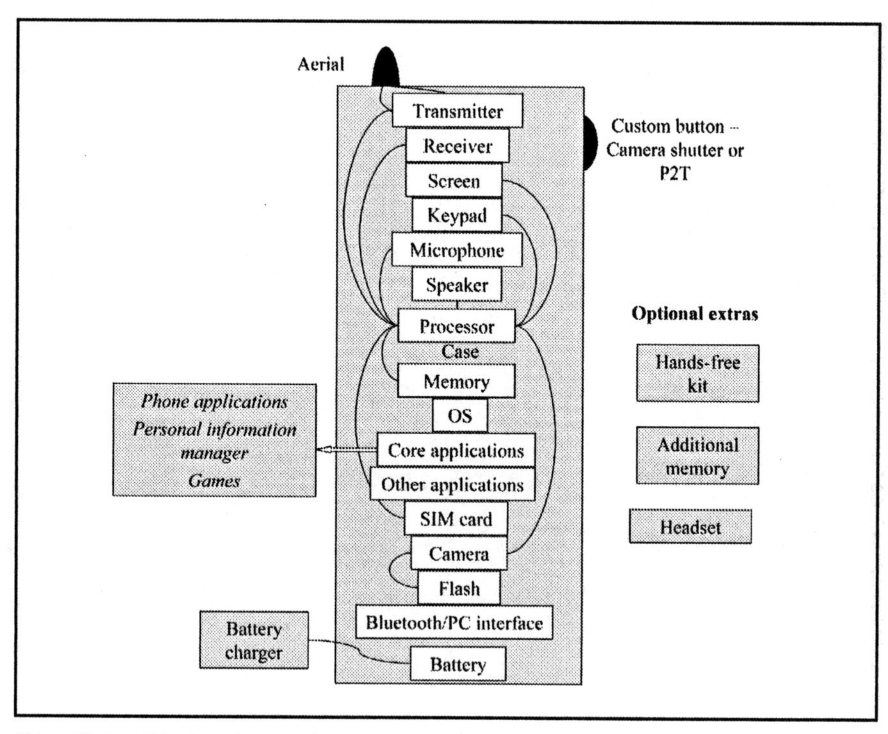

Figure 19. *A mobile phone is a complex system in its own right.*

MP3s (music based), a camera (image centred) and also runs a wide variety of Java games (games centred).

Voice centred	Simple keypad and comfortable to hold when talking
Message centred	QWERTY keyboard. Easy to synchronise with PC
PDA centred	Stylus-based input and large screen, PDA type operating system (Symbian, PalmOS, Windows CE)
Music centred	FM radio, stereo sound with headphones, MP3 player
Image centred	Built-in camera with flash and shutter key, large screen and memory
Games centred	Games-based keypad, fast processor and graphics, large screen and memory

Figure 20. *Six possible ways of designing a usage-based mobile phone.*

Running a mobile operator

When the mobile business began, all the operators had their own networks[4]. Although many of them were simply offshoots of the major and often publicly-owned, fixed-line telecommunications companies this was primarily a start-up market. It was fuelled by new entrants with ambitions to build new telecommunications brands. On the other hand, the fixed-line businesses felt compelled to enter the mobile field because of its potential to reduce fixed-line revenues. Worried by cannibalisation, many of them did not promote their products as aggressively as the start-ups, and hence the giants of mobile are mostly businesses that did not exist twenty years ago.

Becoming a mobile network operator (**MNO**) is frequently seen as an attractive proposition because of its position in the value chain. This means that the operator takes a large share of all revenues, has the most direct relationship with the customer, and still has significant growth potential as the market matures. It would still be possible to start a new network and there are a few examples of businesses that have come late to the game and successfully started a new MNO, such as Nextel (see case study below) and Hutchinson. Nevertheless, the amount of capital required, and the complexities of the task are both formidable.

Increasing value →

Content	Application provision	Service provision	Network	Service access
Video	Video streaming	ISP	GSM	Mobile phone
Music	Content distribution	CPS	3G	PC
Games	VoIP	Network services	PSTN	PDA
Personalisation			Cable	
			Satellite	
			Hardware	
			Software	
			Support systems	

Source: Ofcom

Figure 21. *An indication of how various businesses add value.*

[4] This is not always the case today. See Chapter 5 "Who needs a network?"

Case Study: Push-to-talk and Nextel, a new network operator

The popularity of CB radios in the 1970s (discussed in Chapter 3) provides an interesting pointer to the success of Nextel. In 2003, Nextel launched 'push-to-talk' (P2T) in the US. This 'walkie-talkie' service was immediately popular with users and was seen as a replacement for CB. Because P2T is economical with network resources, it can be sold more cheaply than conventional mobile calls without affecting profitability. Nextel differentiated itself by being first to market with P2T. The company estimates that its customers make 50 % more P2T calls than normal phone calls. Interestingly they offer unlimited free P2T together with a fixed number of minutes on the phone. Thus, in effect, the P2T calls come free of charge. This does in itself mean that large numbers of people will make P2T calls, rather than making conventional telephone calls and thereby significantly reducing revenues to the phone operators. In some important ways, P2T is inferior to mobile phones.

Unlike a mobile phone, P2T does not set a dedicated circuit for the call but runs as a permanent background activity. Users can enter and exit by means of a dedicated button on their handsets; there is no need to dial a number and it provides a faster connection. It operates in a similar manner to CB radios, where users make contact on any selected channel, but transmissions can be made in only one direction at a time. Just one user can speak at a time; all others can only listen until the speaker stops talking. This requires users to manage who has the right to talk at any particular time by saying 'over' to invite others to reply.

Other companies followed Nextel's lead in the US and by 2004, Orange had introduced a P2T service in the UK called Talk Now. Other European and Asian operators quickly followed suit. Talk Now includes advanced features such as group messaging, and conference calling and has moved well beyond the original P2T concept. It is also priced very differently from the US model. Talk Now was launched with an introductory promotional tariff offering users the instant connection and one-to-many feature at the same cost as conventional one-to-one GSM calls. Thus P2T in the UK is less economic for users than in the US and therefore less likely to cannibalise GSM calls or text messages. User trials suggest that Talk Now will barely get any use as a replacement for conventional one-to-one phone calls. P2T also highlights the need for some form of presence management such as the ability to block all communications except those from say your partner, your PA and your boss.

It is clear that P2T addresses some of the same needs as the latest IP communications services such as Windows Messenger. Both services provide rapid informal communications, especially between those working in teams. The clear difference is that P2T is a mobile solution while IP communications require fixed networks accessed through PCs or IP phones. Furthermore, IP communications are significantly cheaper than those using conventional line telephones. Notably, while Nextel's P2T is significantly cheaper than making a mobile phone call, this is not true of Orange's solution in the UK.

What an MNO does

Even in the smallest countries, MNOs employ thousands of staff, with the majority involved in the retail, rolling out and management of the network, and providing customer care. However, they also need to be involved in handset and content provision, billing and marketing their products for business or personal use. Of all these functions, rolling-out and running the network are the only specialised functions that are unique to MNOs.

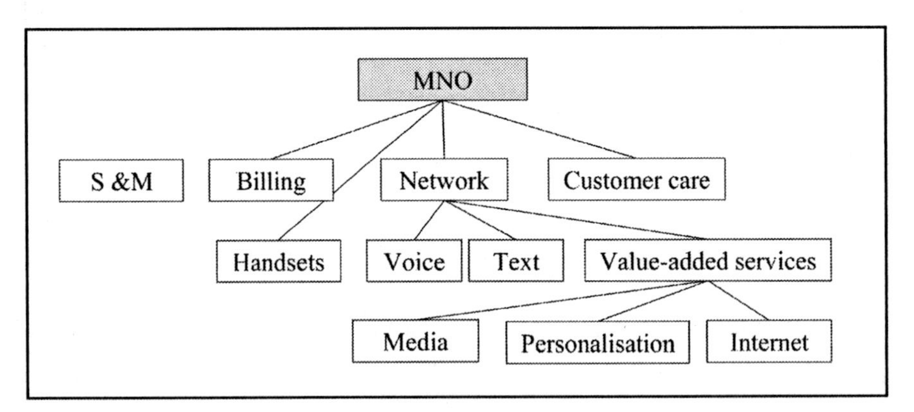

Figure 22. *A summary of the range of tasks undertaken by MNOs.*

Even in a relatively small country like the UK, there is an enormous amount of work involved in setting up the radio network. Sites have to be selected for masts to give continuous cover and then permission obtained from both local authorities and the owners of the land to set up masts. Of course, there also needs to be electrical power available. Then there is the control centre to handle both voice and text messages. Contracts will have to be let for the civil works and all the necessary radio equipment, and the contractors must be managed with care. In addition, handsets must be provided to customers. The operator can choose whether to supply directly, or through retailers and a decision must be made about whether to subsidise handsets and if so, to what extent. The same issues apply to the SIM (subscriber identity module) cards; an essential part of every mobile phone. All of this work has eventually to be paid for by the mobile network users and Figure 23 indicates what phone users are charged for, and how the cash flows among the companies involved.

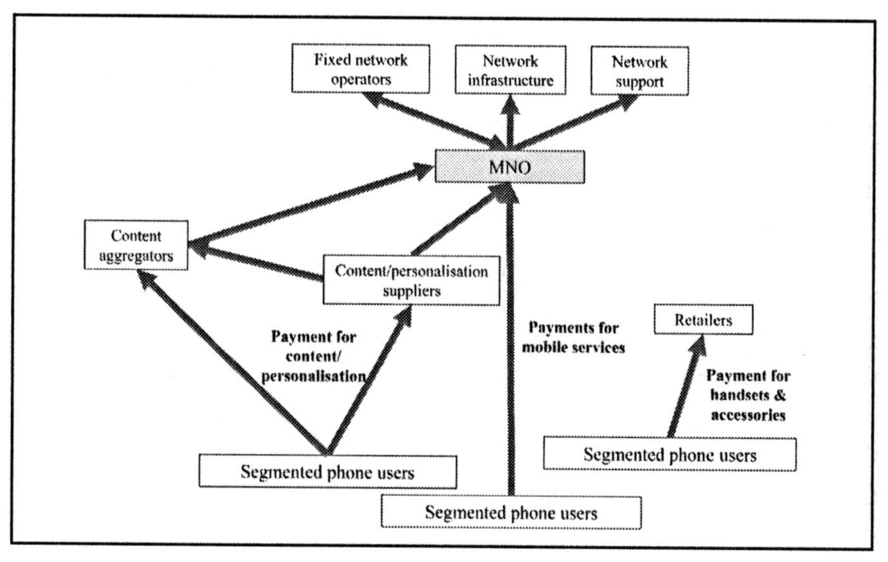

Figure 23. *How the users' cash is spent.*

Then the network has to be commissioned and customers connected. The marketing of a mobile network is a specialist and expensive operation. Consideration must be given to product differentiation, the market segments that will be tackled and appropriate advertising campaigns launched and kept running to win and maintain the customer base.

There are two essential questions for the management of any MNO:

1. How will customers actually obtain their handsets?

2. What pricing strategies will be employed?

Decisions will need to be made about how to set up retail outlets for the brand. The success of a company in what is a very competitive marketplace depends to a large extent on the use of pricing plans that will prove attractive to potential customers – and retain them for the network.

Contractual agreements will be needed with other operators, including virtual operators, covering aspects such as termination charges, roving and the handling of text messages. Similar contracts will be necessary with content providers and details of these are given in subsequent chapters.

Billing requires an efficient financial system to ensure that revenues are gathered in a timely manner and a call centre is the norm for providing customer support. This latter will involve significant staff training as well as an understanding of the foibles both of the network and the various compatible handsets.

A website and a WAP portal are essential tools that customers can access and an effective and ongoing marketing campaign is crucial to attract customers and retain them, supported by appropriate advertising. Distribution aspects cover the need to make sales to customers directly from these sites.

Then the operator may wish to obtain content from premium brands to sell to their customers. This may be done either directly or through a content aggregator to simplify the interface with the growing numbers of content developers. There are also network-equipment supply companies. These companies will provide product-related services such as maintenance, repair and replacement, network building, optimisation and security, turnkey implementation and system integration

Finally, assuming the MNO has set up on a national basis, and almost all have started that way, it will become essential to think about expanding internationally. Perhaps this could be achieved by acquiring or merging with an organisation operating in an adjacent field such as fixed-line or cable telecommunications, Internet service provision or other media area.

The economics of running a mobile operator

Even with all of the complexities involved, many people have argued that running a mobile operator is a licence to print money, and there are some analogies in the way a mobile operator works. In both businesses the majority of cash is spent upfront (on the printing press; on the network infrastructure), and there are very low variable costs compared to the high overheads (paper, ink, customer care, handsets subsidies). However, the difference is that there is not, generally speaking, a legal market for printing money, and the margins of mobile operators are pressured in the same way as any other business operating in a highly competitive market. The economics are as follows:

1. Invest billions of pounds (or dollars or euros) in acquiring a licence and rolling out a network, plus hundreds of millions of pounds annually in running the network.
2. Spend between £40 and £200 ($ or € × 1.5) per customer to acquire them. This includes sales costs, marketing costs, and the cost of subsidies to the handsets.
3. Charge each user between £240-£480 per year for the service.

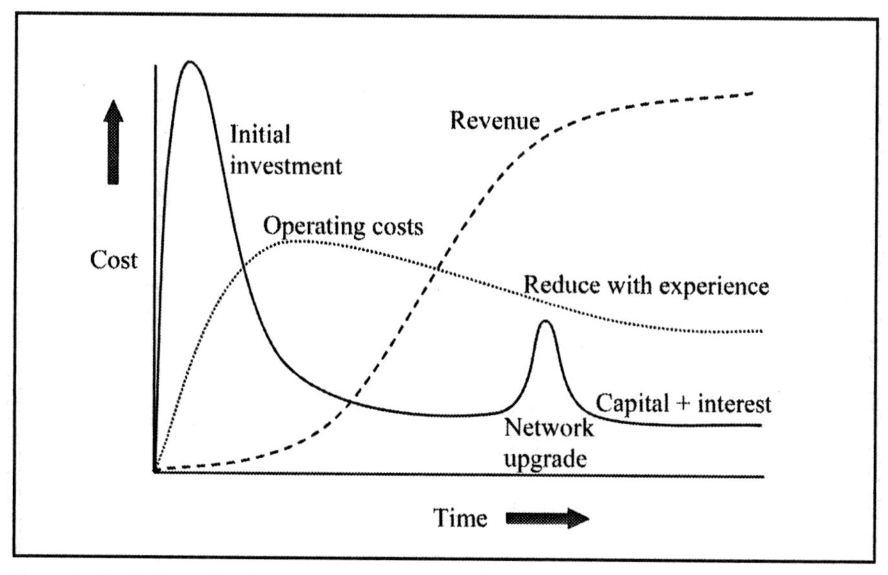

Figure 24. *Typical cost/time investment/revenue profiles for a growing MNO.*

Given that customers are typically retained for a period of eighteen months, you have an average income per customer of £600, with a customer acquisition cost of, say, £100. This gives £500 gross margin per customer. You therefore need millions of customers to cover your annual running costs, and closer to ten million customers to begin being able to obtain a return on your initial investment on the network. A licence to print money, perhaps, but a very expensive licence indeed.

There are now strong consumer brands in the MNO sector, such as Vodafone, T-Mobile and Telefonica, and these global brands are able to produce economies of scale across their business. This has a significant impact in terms not only of purchasing power with equipment and handset suppliers, but also in marketing terms. Vodafone had a strong relationship with Manchester United during their very successful years and benefited globally from the football club's success. Additional sponsorship with David Beckham, their leading striker, provided a 'Vodafone face' across the world that simply would not have made sense to any smaller brand.

At the same time, as the mobile phone market is reaching saturation in many countries, a period of price competition is ensuing that will certainly benefit customers but is likely to leave smaller local operators with a limited future. In the UK, for example, there are no independent operators: Vodafone has become a leviathan in its own right; BT's O_2 has been sold to Telefonica; Orange was sold to France Telecom; and One 2 One became part of T-Mobile. Mergers rather than start-ups are likely to dominate future strategy.

To give an idea of the size of the MNO business in the UK, the revenues of these four companies in the second quarter alone of 2005 are shown in the table below, together with their relative ARPUs (average revenue per user), remembering that each of the networks has a similar number of customers (some fourteen million):

Company	**Revenue**	**ARPU per month**
Vodafone	£2,568 million	£24.9
O_2	£1,848 million	£22.6
<<Orange>>	£1,962 million	£22.5
T-Mobile	£1,380 million	£19.0

Figure 25. *The revenues and ARPUs of the main UK MNOs.*

'3', formerly Hutchinson 3G, is the most recent MNO in the UK market launching a brand-new 3G network. Because of this and its present low market share, its revenues and ARPU are excluded from Figure 25.

MNO revenues

The initial investment in setting up a mobile network is very large and it is not helped by high initial-operating costs that only reduce as operating experience is gained. Revenues, on the other hand, are slow to grow in the early stages and the increase will tail off later as market saturation is reached. Furthermore, as Figure 26 shows, any mobile network operator has to interact with a wide range of other organisations.

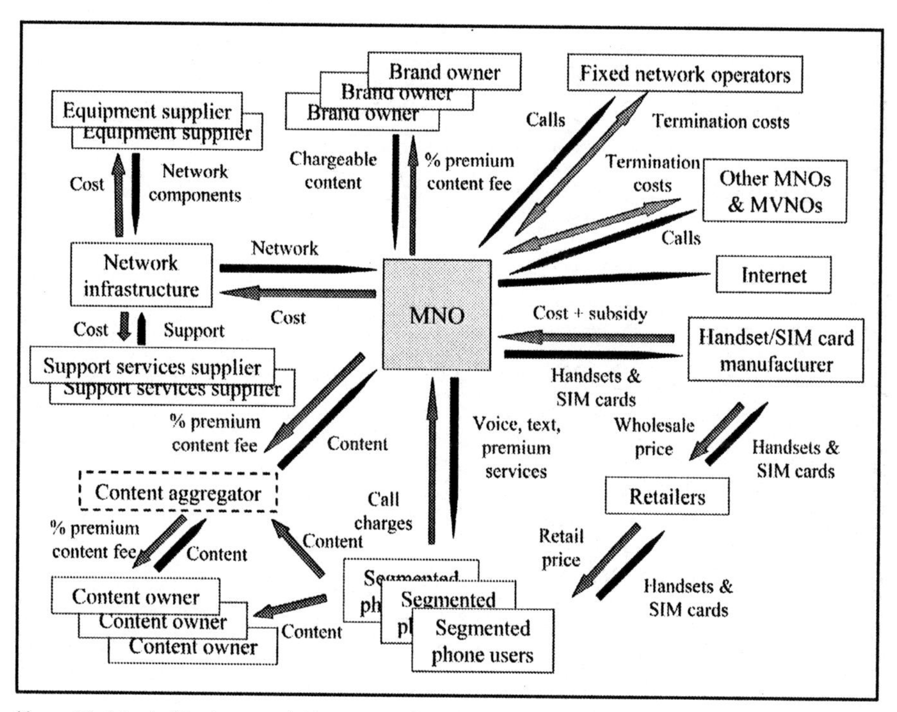

Figure 26. *A typical business model for a network operator showing deliverables and money flow.*

The first mobile phone business model was to invest in a network and to charge the user for the phone itself and then for each call made. Subsequently a number of alternatives evolved including subsidising the cost of the relatively-expensive handset, offering a low price for a fixed amount of call time, a range of tariffs depending on the time, day and distance involved, and pay-as-you-go. The last an attempt to mitigate the cost impact of losing a mobile or having it stolen.

There are many different ways of paying for mobile phone use. The most popular ones are based on one of three concepts: pre-pay, pay-as-you-go and subscription for a given amount of airtime. In all three cases, there are likely to be different prices for text messages, text messages to premium lines, voice calls, calls to premium-rate lines and Internet access (in particular downloads). Costs also vary depending on the hour of the day, whether it is a weekday or a weekend, whether to a mobile on the same network or not and whether the call is to an overseas number or made from abroad.

Payments for pay-as-you-go and subscription service will normally be billed monthly and paid by credit card or direct debit. Pre-pay, on the other hand, allows top-up by text, voice, online, at an ATM or using cash or a card in a suitable store. Pay-as-you-go has primarily appealed to business or professional users who are financially secure, and can afford the additional bill each month. Pre-pay appeals to younger audiences, those who may not have bank accounts and people living closer to the breadline. The older generation often favours pre-pay because they reserve their mobile phones for emergency use only. There is also significant take-up of pre-pay amongst criminals who use it both to make almost untraceable calls, and for money[5] laundering.

Because the market is so competitive, mobile operators have put significant effort into designing a range of price-plans that will appeal to as broad a range of users as possible. Unfortunately, in most countries this has developed to the point that operators often have literally hundreds of different plans, and far from helping the customer find the best plan for them, it simply bewilders people with the array of choice. The table below gives an idea of the wide range of different pricing options available from the UK's largest mobile operator, Vodafone. In addition, there are eight different top-up options.

5. The classic pre-pay laundering scheme is to set up a premium rate number from a legitimate business, typically adult, and then buy pre-pay cards with the illegitimate cash and use the cards to call the premium rate number.

It can come as a surprise to no one that potential users find it difficult to understand complicated offers, which are frequently further confused with small-print additions. It's no wonder that easy-to-understand price options are proving popular.

Anytime	Mobile Alert & Mobile Alert Plus
Evening & weekend	MEPOS
Pay as you talk	Vodafone Mobile Connect card *
Smartplus	**For groups**
Smartstep	Sharetime 2500
Text packs	Sharetime 8000
Voice packs	Businesstime 1200
In-store pay monthly price plans	Businesstime 800
Pay monthly offers	Sharetime 24000
Extra value plans	Sharetime 750
Pre-pay service	Businesstime
Pay-as-you-talk	Businesstime 450
Online exclusive price plans	**For one man bands**
Satellite phones	Businesstime 300
Insurance	**Paging**
3G price plans	Sharetime 14000
3G/GPRS price plans	Sharetime 5000
3G/GPRS card	**Costs abroad**
GPRS card	International call saver
SIM only	Sharetime 1250
Pay-as-you-talk	Sharetime 500
Pay-as-you-talk offers	Businesstime 200
Pay monthly	Data costs
	* for laptops

Figure 27. *The range of pricing options offered by just one MNO.*

On the marketing front, at the end of 2005 Vodafone was making two offers to its customers. This first stated: 'The latest mobile phones plus half price line rental.' The second offered 'Free Sky Mobile TV – exclusive to Vodafone.' Both included a choice of different free handsets. The offers are illustrated in Figure 28.

Figure 28. *Two concurrent Vodafone promotions in November 2005. (Courtesy Vodafone)*

In addition to revenues from consumers, a significant proportion of a mobile operator's revenues comes from termination charges. These are charges imposed when calls to a mobile phone are made between network operators, whether the call originates from a mobile or fixed phone.

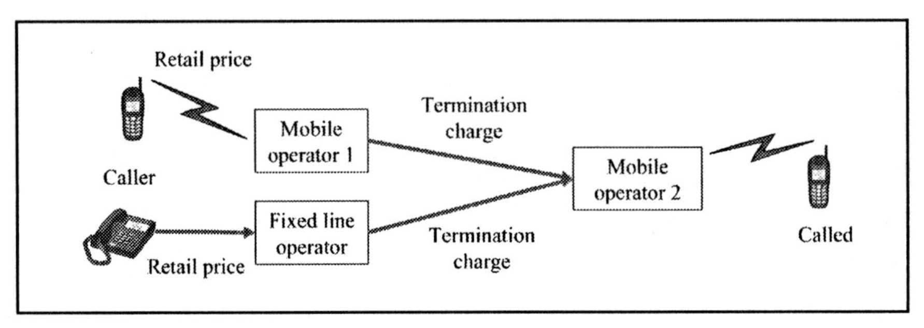

Figure 29. *The way that call termination charges are imposed.*

There has been some controversy over the variation, amount and control of these charges and there are a number of regulatory investigations into the pricing level of call termination. Since these fees are not immediately apparent to customers, they have not been subject to the same rigours of competition, and particularly in the case of calls between operators in different countries, the fees are frequently excessive.

The mobile ecosystem

There are plenty of opportunities to work for an MNO. Apart from the internal roles of sales, marketing, managing suppliers, and letting contracts, MNOs require billing, customer care, retailing and distribution as well as a website and WAP portal. While any or all of these tasks may be undertaken in-house, there are significant openings to supply these services as a sub-contractor. Many businesses have set themselves up to provide facilities such as call centres, retail outlets, billing systems and Internet site designs and maintenance.

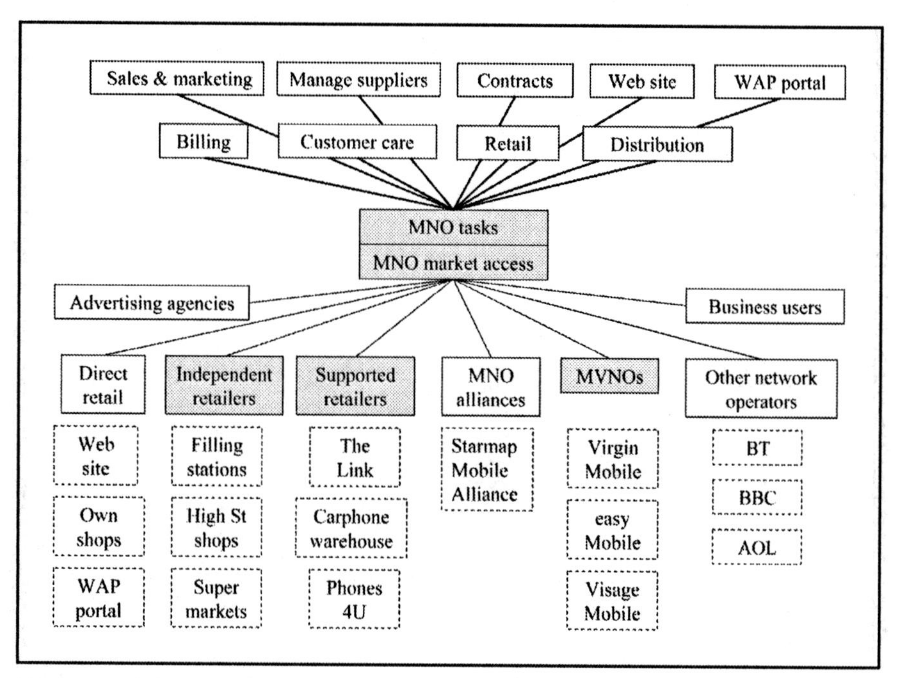

Figure 30. *Both the tasks undertaken by an MNO and its marketing options suggest other business opportunities.*

Case Study: Vodafone

In 1982, Racal Electronics Group bid for and won the private-sector UK mobile licence and set up Racal Telecomms with fewer than fifty employees. The Vodafone analogue mobile phone (TACS) network, the first in the UK, was launched early in 1985. Its rapid roll out quickly achieved the required licence population of nineteen thousand within a year.

Vodata, created in 1987, was the 'voice and data' business to develop and market Vodafone's voicemail service. Other services were added, such as the first information lines (Financial Times and AA Roadwatch). Vodapage was also launched, providing a paging network covering 80% of the UK population. Committed to aggressive long-term development of international mobile telephony, the company joined consortia bidding for licences, and identified acquisition opportunities.

By 1988, the business provided a third of Racal's annual profits. The following year Paknet, a joint venture between Racal Telecom and Cable & Wireless, was formed to meet the business world's needs for sending fast, reliable and accurate packet data in fixed and mobile applications. In addition, a back-up network was introduced to reduce the effect of any system faults.

In 1991, the Vodafone Group became independent following the largest demerger in UK corporate history. In the same year, two milestones were the world's first international call and the inauguration of the first digital (GSM) mobile phone service in the UK.

The early 1990s saw a range of new tariffs for consumers, the first Vodafone high street outlet and a distribution agreement with Comet, a major UK retailer. At the same time, its number of international partnerships grew significantly and Vodafone Group International was formed to acquire licences and supervise overseas interests.

In the mid-1990s Vodata became the first network operator in the UK to launch data and SMS services over the digital network. Vodafone also pioneered a pre-pay analogue package, while per-second billing was introduced together with options to purchase 'bundled' minutes and make off-peak local calls to fixed-line phones. This was shortly followed by the digital 'Pay-as-you-talk' package. In addition, its six wholly-owned service providers were reduced to three, allowing it to rationalise its billing and customer-care systems, and streamline its brands and tariffs.

In 1999, Vodafone Group and the US AirTouch Communications Inc. merged to create Vodafone AirTouch plc, the world's largest mobile communications company and one of the top twenty five companies of any kind, worldwide. Its customer base exceeded thirty million.

The new Millennium started with Vodafone's obtaining the largest available UK 3G licence and acquiring Mannesmann AG (despite having to sell Orange to France Telecom), almost doubling the size of Vodafone. In 2000 Vizzavi, a 50/50 joint venture with VivendiNet, launched a multiaccess branded Internet portal for Europe. While Europe continued to be the largest part of the company's operations, the Americas and the Far East provided a growing customer base.

2002 saw Vodafone trial its global mobile-payment system in the UK, Italy and Germany to enable customers to purchase physical and digital goods using their mobile phone. It also chose Ericsson as a global MMS supplier and Siemens as a global location-enabling server supplier to allow Vodafone to create new applications. Vodafone and T-Mobile created an interoperable mobile-payment platform and Vodafone launched the first commercial European GPRS roaming service. This facility allowed customers to seamlessly access services such as corporate e-mail, intranet and personalised information on their mobile phones – or their laptops, pocket PCs and PDAs. Later that year, Vodafone announced Vodafone live! for consumers and Mobile Office for business users, quickly followed by Vodafone Remote Access, part of Mobile Office that gave travelling customers easy access to their corporate IT networks, to access e-mail, calendar and other business specific information. Vodafone also signed global agreements with many of the leading computer equipment suppliers to offer business customers an extensive range of 'Connected by Vodafone' mobile-computing devices. This included notebooks, pocket PCs, PDAs and tablet PCs with easy connection to Vodafone's GPRS data network.

Conclusions

Since 2001 the telecommunications industry has experienced tough conditions, on a global level. The problems started in the equipment part of the wired-network industry and then spread to the mobile equipment and eventually to the service segments. In the manufacturing sector significant job cuts occurred within large multinational companies. It seems clear that as the use of mobile phones has grown, so has the cover and competition provided by mobile network operators.

1. The global mobile phone business has mushroomed since the Millennium. It is worth some $600 billion per year. This presents many openings for you to grab a share of these huge revenues.
2. The mobile phone business is huge and complex with intense rivalry between the now very large international operators.
3. It would be unwise to launch yet another MNO.
4. There are plenty of opportunities to act as suppliers to MNOs and many are described in subsequent chapters.

Chapter 5
Who needs a network?

"So how long is it before Madonna sets up an MVNO, selling die-hard fans phones with Madonna-ized faceplates, wallpapers and ringtones, with messages or voice greetings from the star." - **Carlo Longino**

Introduction

By the end of 2005, the established UK mobile business was worth £10 billion in annual revenues from consumer services alone. This is big business. Recent primetime viewing on ITV was The X-Factor, a talent show that drew more than six million viewers and provided its broadcaster with some of the most expensive advertising space it could sell. Every single advertisement was for a mobile phone company. Nokia sponsored the show itself and are still promoting the 2006 X-Factor tour on their website. The mobile phone companies are buying the best advertising slots because they are making enormous sums of money.

At the same time, annual surveys of the most powerful brands rarely include mobile phone companies (Vodafone occasionally features), so it beggars the question: 'Can major brands make money without investing the capital to build a network?'

For many consumer brands, launching a mobile service can have a significant impact on their bottom line figures. There are three reasons why most brands have not to date launched an MVNO (Mobile Virtual Network Operator) operation:

1. It is a major endeavour – There are high costs for establishing the operation; up to £10 million per year, and most of this money is not recouped until the network reaches some half-a-million subscribers.

2. You need significant competence to succeed – There is significant up-front risk for a business that does not have expertise in operating a mobile network. There is a world of difference between most retail or consumer goods businesses, and the capital intensive nature of managing a mobile network.

3. Launching an MVNO requires support from a network operator – Most operators have been reluctant while there have still been significant new customers for them to win. Network operators have been turning away wholesale deals for the last few years and even now, they are unlikely to be equipped to handle the number of potential deals on offer.

However, all of these factors are mitigated by the argument that it makes good business sense for a premium brand to operate a phone network. Barriers of scale and expertise are being lowered by the emergence of Mobile Virtual Network Enablers (MVNEs), providing many of the services required for preparing the launch of a virtual network. Whilst the market in Western Europe reaches saturation, many operators are all too happy to buy market share by piggy-backing new brands on their networks, especially as one of the benefits for the network operator is that there is zero cost for additional minutes sold.

Ultimately, an explosion is likely in the number of MVNOs over the next ten years. For any major brand owner, this is a great way to profit from the mobile revolution.

When to use an MVNO

An MVNO is set up by an organisation that has the scale to market and distribute mobile devices but, for financial or operational reasons, does not want to invest in setting up a radio network. By making a wholesale deal with an MNO to provide network services, the virtual operator only needs to set up and operate their core business. The MVNO's customers will buy their mobile phones from the brand, and walk around all day long with your brand in their pocket, while the host network runs the day-to-day operation of the business.

The first virtual networks were launched in the late 1990s, and over the last five years more than one hundred and fifty organisations have started offering virtual mobile phone networks in over twenty-five different countries; mostly in Europe. MVNOs initially came out of the operator's desire to avoid the sort of anti-competition regulation that forced conventional telcos to provide indirect-access services (see Chapter 2), but also as a low-cost form of brand segmentation.

When you are looking to start an MVNO, you must understand that any new provider of mobile services brings more intense competition for the mobile operators. Operators will only host services for MVNOs that can help them target new segments without the significant costs associated with service customisation. Secondly, unless you are working with a very cost-effective network enabler, you will need to be able to achieve five hundred thousand or more subscribers to make your services profitable.

When considering the customers you will be targeting, it is important to make sure that they will provide you with a level of revenue (ARPU) to match your expectations. The usual way of segmenting the market is by disposable income and mobile phone usage. This is demonstrated in Figure 31, which shows four different groups, each representing about a quarter of the market by quantity (not profitability or potential).

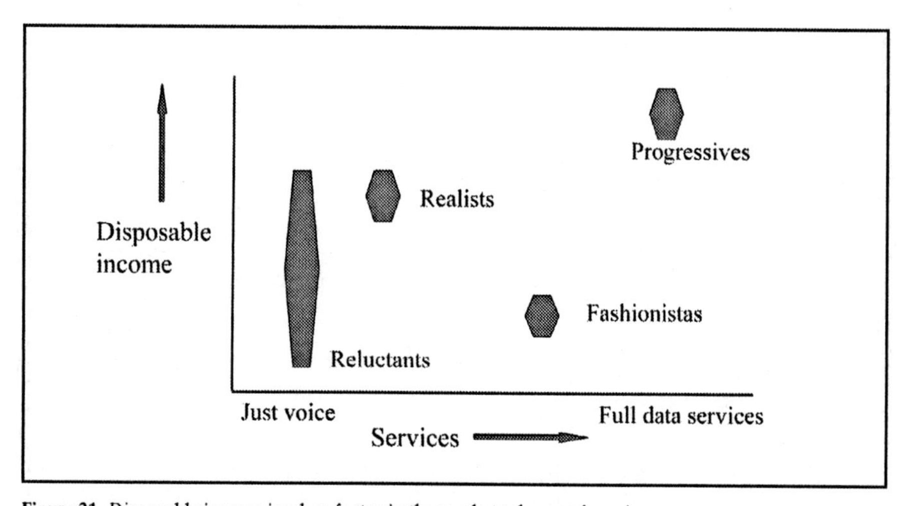

Figure 31. *Disposable income is a key factor in the market take-up of services.*

The four segments, based on attitudes, are:

- Fashionistas – The young market that is focused on fun and community. Mobiles are a part of their life and they bond strongly with their devices. They are early adopters but with limited funds. They are uninterested in professional services. People in this segment should mature to become progressives.
- Progressives – Ambitious early adopters with disposable income available for new services who focus equally on work and leisure.
- Realists – Career and family focused. They can spend but prioritise on using mobiles for practical purposes and for security. They tend to be uninterested in enhanced data services and content.
- Reluctants – The low usage group – enhanced data services do not appeal and the mobile is used mainly for emergencies.

By far the most profitable segments are Fashionistas and Progressives. Your MVNO is most likely to succeed if you can easily reach a large number of these people. Conversely if most of your customers are Reluctants or Realists then you will probably not have much success with a virtual network. It would not, for example, be a good idea for Saga (the group selling to the over fifties) to launch an MVNO unless they had an effective strategy to change their customers from being Reluctants to Progressives. This might be possible by competing on cost or bundling in other services, but it would still be a hard sell.

Getting a deal with a mobile operator

When approaching a mobile operator to discuss a wholesale deal, you should consider in which areas they are strong, and target your MVNO towards an area of their weakness. An operator is much more likely to do a deal with you in a segment that it cannot effectively reach itself. Good examples are ethnic minority groups, which often have a high proportion of Fashionistas and overseas callers. They can be easily targeted by separate brands with own-language billing and they can relate to promotions by community-relevant celebrities.

What an MVNO does

The first question to consider when you are planning to start an MVNO is how much work do you want to do yourself? If the value of your business is in your brand or distribution, then you are probably better off focusing on these aspects and leaving the running of the mobile phone network to people who specialise in that area. In this case you would be running what is termed a 'thin' MVNO using the host network operator's basic infrastructure, and you would provide little more than billing and customer care services. Your wholesale contract with your operator would cover a full range of services and you would only really be differentiating your service on brand and distribution.

If on the other hand, you have a degree of expertise in mobile telephony, or believe that you can run better or cheaper services than your MNO, you should run a 'thick' MVNO. In other words you would be building out your own core infrastructure, messaging platforms and intelligent network so that you can offer your own comprehensive services. In this case, the wholesale contract with the MNO would be just for airtime over the radio network.

Your choice of thick versus thin will have a significant impact not only on the amount of time and effort that you need to invest into establishing your MVNO, but also the extent to which you will be able to provide customised services to users.

1. Thin MVNOs will normally source their own billing and customer care services. They will establish their own tariffs, customer care mechanisms, and provide their customers with basic voice/text services, usually with simple Internet access and some premium content services. They will not be able to provide custom facilities like voicemail, advanced messaging (MMS or SMS) services, or offer advanced Internet or e-mail services separately from those of their host operator.

2. Thick MVNOs, in addition to their own billing and customer care, will frequently run some of their own network services, typically SMS or MMS, voicemail, overseas-call routing, and WAP gateways. This allows these virtual operators to provide advanced services such as VoIP (Voice over Internet Protocol) routing for overseas calls (see Chapter 10), more sophisticated e-mail products (see Chapter 6), or custom-built handsets.

If your business does not have significant expertise in mobile telephony then you should almost certainly use a Mobile Virtual Network Enabler or MVNE to run many of these services for you. Most MVNEs offer a number, if not all, of the many services that a virtual operator requires. These include tasks like administration, provision of operation and business support systems and access to a network. The virtual operator normally pays the enabler a pre-arranged percentage of its revenues, while some enablers even have an agreed wholesale rate with potential host network operators.

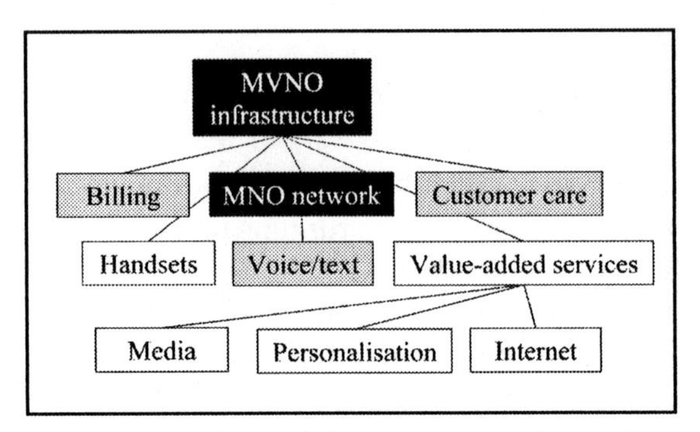

Figure 32. *A typical MVNO infrastructure showing in grey what a thin one might provide and the additional features (in white) that a thick MVNO could offer.*

The MVNE provides a hybrid model as many of the components of the thick MVNO may be offered to you by the MVNE and you can therefore become significantly less reliant on your host operator, but will of course depend more on your chosen MVNE.

Regardless of whether you adopt a thick or thin approach, you are going to need to build up a new team to run your MVNO. Most of the sales and marketing expertise for your channels should already exist in your business – otherwise you shouldn't be starting an MVNO – and most of the new activities will be related to price planning, specifying consumer propositions for mobiles, and arranging cross-promotions with your other product lines.

The operations side of the business is significantly more complex, and should focus on four key supplier areas:

1. Network services – In addition to managing the supplier relationship with the network operator, there will be significant technical integration issues to provide services such as rating, billing, customer care and e-commerce.

2. Handsets – Remember handsets very quickly go out of date and must be compatible with your chosen MNO's network. This is not only a matter of keeping up-to-date with customers' desires and upgrades to your operator's network. Don't forget you will only be able to purchase any given handset for a limited time period. Managing your stock levels will have a major impact on the profitability of your business. You will need to judge carefully which handsets will be top-sellers, exactly when you should launch them and whether you should subsidise their retail price. Remember that a lot of your cost lies in the provision of handsets.

3. Customer care – Customers must be able to check their bills, and query problems with the services through a high quality of telephone-based support. If you are starting a price-led MVNO you could provide a lower quality of service, perhaps limited to self-service over the Internet.

4. Content services – Ringtones, text alerts and premium content can take up a significant amount of management time for little revenue. Chapter 7 provides more details on this area but, needless to say, unless you are operating a strategy that requires significant content play, it would be better for you to completely outsource this to a third party. In addition, network operators might be prepared to engage more readily in the 3G MVNO opportunities as a way to reduce the cost burden currently associated with poor 3G take up.

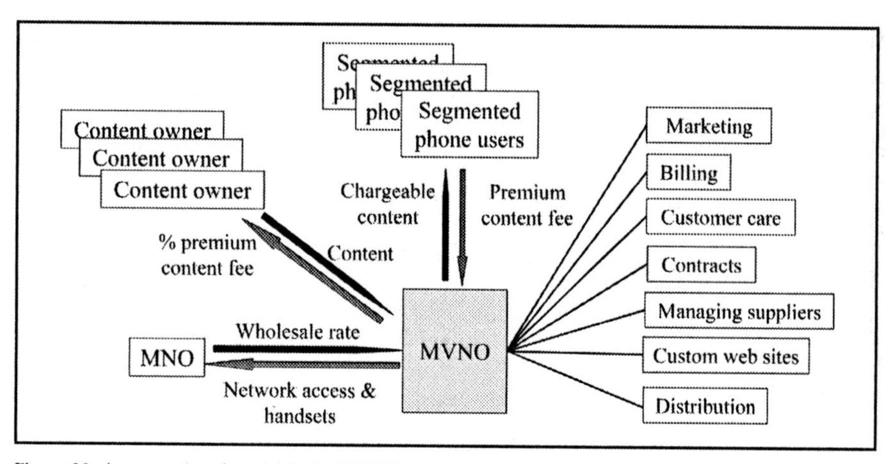

Figure 33. *An operational model for an MVNO.*

MVNO strategies

Most mobile network operators describe competition in their markets as ferocious. This is particularly so in Europe where price per minute is high enough to maintain good margins even when mobile networks are operating nowhere near their full capacity. Most operators now realise that selling airtime to MVNOs can increase their revenues, but at the same time they consider that virtual operators are a significant part of the reason why the environment in which they function is so aggressive.

They are anxious to avoid MVNOs eating into the high-end customer base by differentiating themselves on quality of service. Most MVNOs are left with a relatively basic service compared with the incumbent operators and often have new handsets a long time after their host[6] operator has launched them.

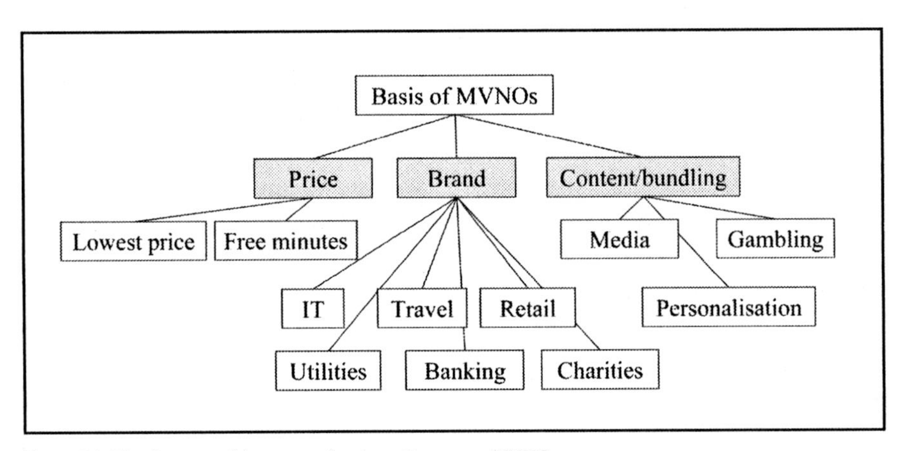

Figure 34. *The three possible approaches to setting up an MVNO.*

You can take one of three basic approaches to running an MVNO:

- Build it on a well known brand.
- Establish a price-winning formula.
- Bundle together premium content or other services.

[6] Given the quality problems associated with the launch of new handsets, this can of course be a blessing in disguise for the MVNO.

Of the three strategies, bundling is probably the most attractive to many businesses, but it is also the most risky as it has yet to be proven on a large scale. You will also need to select your target market segment very carefully.

Branded

Brand and brand loyalty are crucially important to a company's success in selling to the mass consumer market. Many companies have exploited their brand image to make successful moves into new markets. Richard Branson's Virgin empire has expanded from selling records to running an airline. Oil companies have started selling groceries through their petrol outlets, and many UK supermarkets now offer personal banking, clothes and electrical goods. It is thus unsurprising to find that a number of successful brands are keen to move into the mobile phone business.

The importance of brand in the mobile sector is perhaps even greater than in other areas, with many of the best marketing executives being lured to join mobile operators. If you are a well-known name in the retail sector – a supermarket, travel agent, high-street bank or department store – or a charity or IT company, then you will probably have sufficient brand leverage to launch an MVNO based on your brand. Alternatively you may have strong distribution but a less well-known brand. The key is that you must be able to sign up sufficient customers on the basis of the strength of your brand or distribution to reach the critical mass required for a successful virtual operator.

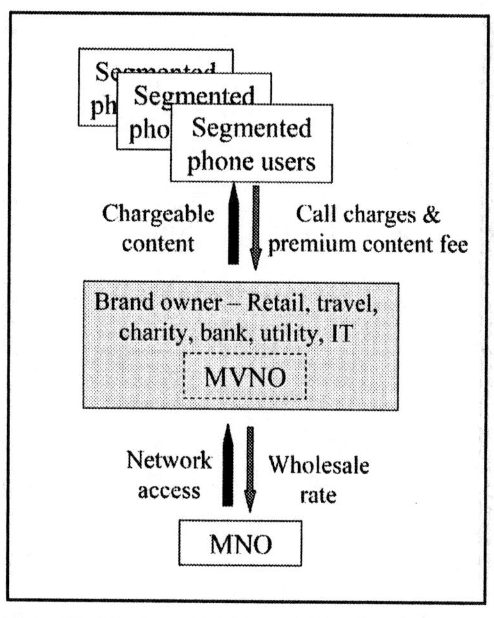

Figure 35. *A business model for an MVNO based on brand.*

So what does the brand offer to customers of its mobile service? Potential consumers are typically the brand conscious people, who strongly identify with the brands they love. Often Fashionistas, they are high-revenue mobile phone users. It naturally follows that they should be given the chance to buy a mobile from their favourite brands at a competitive price. In return they will receive a badged phone with brand livery, sometimes bundled together with relevant value-added services.

The key advantage that the branded MVNO has is the leverage in marketing the mobile services alongside other branded products, the reduced cost of utilising existing distribution channels, and the reduced churn from customer's loyalty to their brands.

Case Study : Virgin Mobile

Virgin Mobile is the classic brand-based MVNO. Started as a joint venture between Virgin and T-Mobile, it launched in the UK at the end of 1999. Within a year, it has signed up half a million brand-loyal customers. Less than one year later that number had doubled. This is fantastic growth even for one of the UK's best loved brands.

The products and services are distributed through some five thousand third-party sales outlets, many general retailers in the UK and Virgin Megastores, as well as through the Internet and Virgin Mobile customer-care centres. Virgin Mobile employs some one thousand four hundred people on four sites in the UK that include an outsourced customer-service centre.

The effect was to improve the perception of the Virgin brand with Virgin Mobile becoming a distinct and well-established brand in its own right. Turnover grew to nearly £½ billion, with operating profits over 15%. Compared to the enormous capital expenditure of rolling out a full network, Virgin Mobile claims one of the highest returns on capital employed within the industry.

It is not just the financials that are impressive. Regular surveys of network quality in the UK show the power of the brand, with Virgin consistently ranked, by consumers, as having the best network and T-Mobile the worst. Of course, T-Mobile runs the network for Virgin, so the difference in the perceived quality appears to be solely down to the differences in branding.

The operation is also considered a great success for T-Mobile, which has the fewest subscribers signed up to its own branded services in the UK, but more subscribers than any other network when the Virgin Mobile numbers are included.

Price

Next on the list of possible strategies comes the price-based MVNO. These virtual operators seek market share by offering the cheapest possible price deal because they can (and do) frequently acquire customers in droves, many customers being driven by the attraction of saving money. Potential areas for cost-saving are in reducing customer support costs, not providing handsets, negotiating special rates for overseas calls, or creative tariffs around peak time-usage.

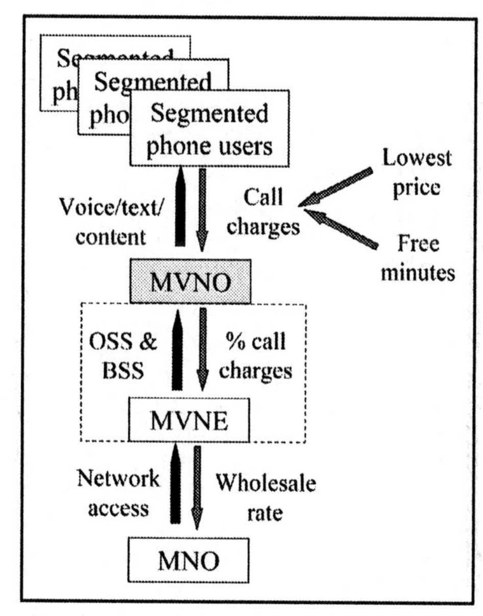

Figure 36. *An example of a model for a cost-conscious MVNO that could also employ an MVNE.*

A common approach is to provide Internet only customer service, where users sign-up for their accounts online, receive their bills by e-mail and query any service problems using an online FAQ. This approach can frequently give a £30-50 cost advantage per subscriber per year compared to conventional distribution and customer service.

A 'no handset' (i.e. buying the SIM card only) approach eliminates the need to provide handsets subsidies, but appears to have limited attraction to customers. Buying a SIM card to put into an old handset for marginally cheaper calls seems less appealing than getting a free new handset. Customers often give their former handsets to older members of their family who use the SIM card for emergencies only – a very unprofitable approach for the MVNO.

Special rates for overseas calls can use VoIP services (see Chapter 10) to reduce the costs of international calls. When targeted at immigrant communities, who frequently make calls to family members in their country of origin, this can be one of the most profitable MVNO strategies.

Additional incentives can be tie-ins with loyalty schemes or tariffs based around times when networks are under-utilised. These are normally targeted at the pre-pay market only. MVNOs offering low-cost, no-frills, pre-paid services that prove attractive to lower-income consumers who cannot afford traditional post-paid services. Even in the wealthy US, some 40% of the population cannot get credit.

The model in Figure 36 shows the optional employment of an MVNE. This frequently provides the lowest-cost solution. Everything in this business model must be geared to keeping customer charges as low as possible and making offers that attract new customers.

Case Study: Telenor and easyMobile

Perhaps the first example of a price-based MVNO occurred in Denmark, following the Norwegian company, Telenor's entry into that market. The company's operations provide a good example of how to gain business based on a lowest-price model.

In mid 2000, Telenor (formerly Norwegian Telecommunications) purchased a controlling (later total) share of Sonofon, the pioneer of mobile phones in Denmark, to gain an entry into that market. Early the next year, it launched a GPRS-based service in Denmark and by mid 2002, it was offering a fixed subscription fee, cheap rates per minute and free GPRS. Within three months, the monthly subscription was also free although the rate per minute went up ten percent. The change also included cheap text messages. By the end of that year, it had built up its customer base to one point two five million and profits to the highest yet in the company's twelve year history. In 2004 the company expanded again by purchasing a Danish service provider, showing how a business can flourish by using a price-winning formula. In effect, Telenor had set up a price-based MVNO and challenged its Danish rival, TDC in a price war. The result is that Denmark enjoys some of the lowest prices for mobile phone usage in Europe.

And TDC benefited from the experience, being chosen as partner to possibly the most noteworthy low-cost MVNO in the UK – easyMobile. Stelios Haji-Ioannou, the man who founded the low-cost airline easyJet, set this up. As with his successful airline business, he started an aggressive price war in the mobile phone sector. His unique selling proposition is based on persuading those who already have a mobile phone to switch to a rival operator, in this case easyMobile, by purchasing a £20 SIM card. Customers each receive £20 worth of calls each month charged at low voice and text prices, with even lower three-month introductory rates.

Not only did this remove any premium for 'pay-as-you-go' virtually at a stroke, it also meant that people who do not make much use of their mobiles should gain financially by cancelling their existing monthly contracts. The immediate result of the easyMobile launch was that the Carphone Warehouse and Tesco MVNOs cut their own prices.

However, it is early days to judge the likely success of easyMobile as the company has several hidden costs. These include charges to top-up credit, replace a SIM card, or if customers become overdrawn on their accounts. In addition there is a sizeable cost involved in changing an account name and any problems are only dealt with by e-mail.

Other companies are challenging easyMobile, such as Carphone Warehouse with its 'Fresh' MVNO and the offer of reduced costs for users topping up their phones in the company's high-street outlets. What is clear is that other MVNOs and MNOs have responded to the pay-as-you-go pricing challenge, often with very easily understood tariffs.

Bundling

Finally, there is the MVNO that bundles together mobile with other services. Typically these services, e.g. music or video can be consumed on a mobile device, but there could also be other telecommunications services, such as fixed-line calls or Internet access. To date, most of the experiments in bundling have been by media companies, which have the advantage of appealing primarily to the Fashionistas, but it should be noted that none of these models has been thoroughly proven in the market.

Over the last four years, most mobile operators have started selling content over their networks (see Chapter 7). There is a feeling among many media companies that the large revenue percentage taken by MNOs is unjustified. Some of these companies would like to move up the value chain; several have created their own web portals and a few are now setting up their own MVNOs. They are, however, viewed cautiously by many operators that see potential competition with a strong brand-based offering, particularly from the more aggressive media organisations, such as Rupert Murdoch's empire.

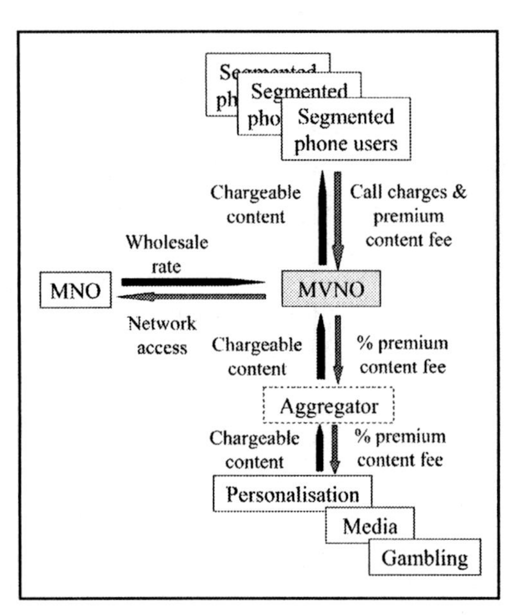

Figure 37. *An MVNO based on bundling may involve an aggregator.*

By bundling media content together with voice and data traffic, a media MVNO can avoid price competition and gain significant customer loyalty without the need for extensive marketing of the products. In some cases, the content services may not even be owned by the media company that starts the MVNO – for example, an FHM mobile would licence much of its content from a content aggregator – and content aggregators or gambling businesses could even launch MVNO services themselves.

There are many good reasons why a media company might want to get involved in providing content to mobile phone users. Of course their content is unique and often part of a strong brand like Disney or EMI. They will already have an extensive customer base and an MVNO will provide a new revenue stream and/or increase their share of any mobile revenues.

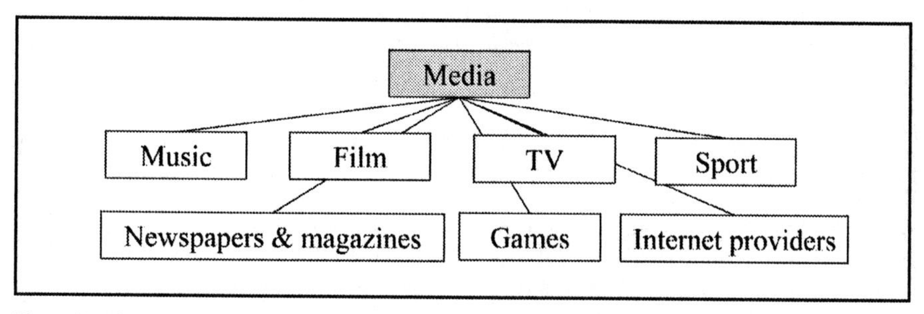

Figure 38. *There are many different types of media companies.*

On the other hand, they are unlikely to have any experience in the telecoms field and will have to invest a significant amount of money in the venture. As most media companies already have existing licensing arrangements, they will have to take care to avoid cannibalising these, and inevitably there may be a danger of their MVNO ending up just selling voice and text.

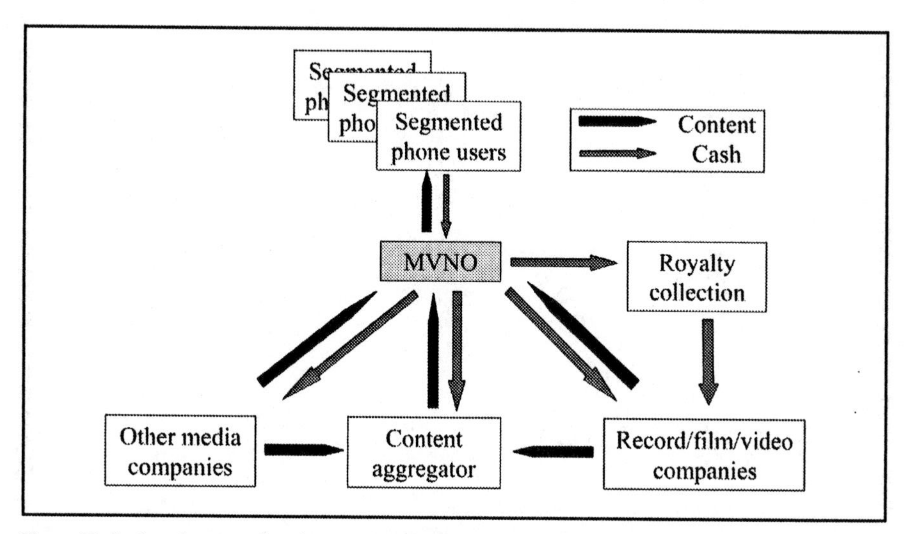

Figure 39. *Cash and content flow for a media MVNO.*

Any radio broadcasting or record business can provide music content, while images can come from film or TV companies as well as music videos. These all have the potential to link their brands with MVNOs, as do cinema chains. Other opportunities are also offered in the sports arena, the clubs themselves, and TV channels providing suitable content. In the games sector, there are possibilities with electronic game suppliers in addition to games console and toy manufacturers. Finally, newspapers and magazines, Internet service providers and even Internet search-engine companies may be suitable candidates.

There are several likely approaches that companies will choose to examine for provision of a bundled MVNO. Figure 40 shows three of these possibilities. The first is a thin MVNO based on a single brand that does not take advantage of the media company's content. It might employ an MVNE to provide operation and business support systems as well as network access. The MVNO would pay the MVNE a pre-arranged percentage of its revenues, while the MVNE would have an agreed wholesale rate with the MNO. The products that could be bundled, in this case, are primarily products that are not consumed on the mobile phone, typically financial products like those offered by Royal Bank of Scotland or American Express.

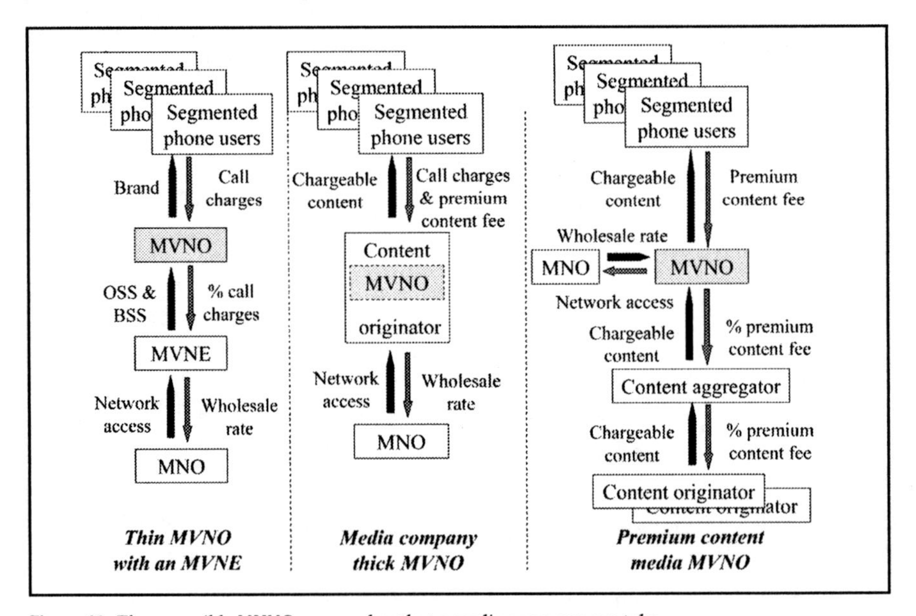

Figure 40. *Three possible MVNO approaches that a media company can take.*

The second choice is an MVNO entirely owned by the content supplier that retains the premium-content fee but pays an MNO an agreed wholesale rate for access to the network. This is the standard media MVNO bundled model and is the route taken by both ESPN and Universal (see Case study: Media MVNOs).

Case study: Media MVNOs

The two examples of media MVNOs to date are Universal Mobile and ESPN Mobile. The former is a successful French MVNO that provides a mass of content to its users. It is a partnership between Universal Music and Bouygues Telecom that was started in mid 2004 to provide a music download/mobile phone brand. Universal Mobile achieved some quarter million customers in its first year of operation.

The American company ESPN is a well-known entertainment company with radio, TV, Internet and sporting interests. ESPN Mobile offers voice as well as unique sports programming and entertainment specially tailored for delivery over an ESPN-branded mobile phone. These services include sporting results, analysis and statistics, special ringtones, graphics images, photos and logos in addition to streaming of audio and video. Its content is aimed at the sports enthusiast, mainly young male professionals. It has chosen to lease network airtime from the Sprint MNO.

An unusual example is MTV Sweden, which has set up MTV Hello, using Spinbox as an MVNE, to gain revenue from voice and text but for a number of reasons it has not taken advantage of its own content. It is interesting to note that MTV has a number of licensing agreements in the US and Europe. It may be that concerns about cannibalisation have prevented it offering premium content, but this case is closer to a media brand offering a branded MVNO rather than bundling, although it is likely in the long term that MTV will pursue a bundling strategy.

In the third model, the MVNO still pays for network access, but also obtains content from an aggregator for a percentage of the premium-rate fee. The aggregator then pays a smaller percentage to the content originators. This model would appeal most to businesses that have a strong brand and a smaller range of content services – such as a gambling or music – but not the full range of content services needed to make the proposition compelling to the user.

MVNO business models

The basis of an MVNO, and the source of most cash flow through the business, is the wholesale deal with the network operator. Regardless of the level of bundling, it would be very unusual for significant revenues to come from other sources, and the basic business model is the same for all strategies.

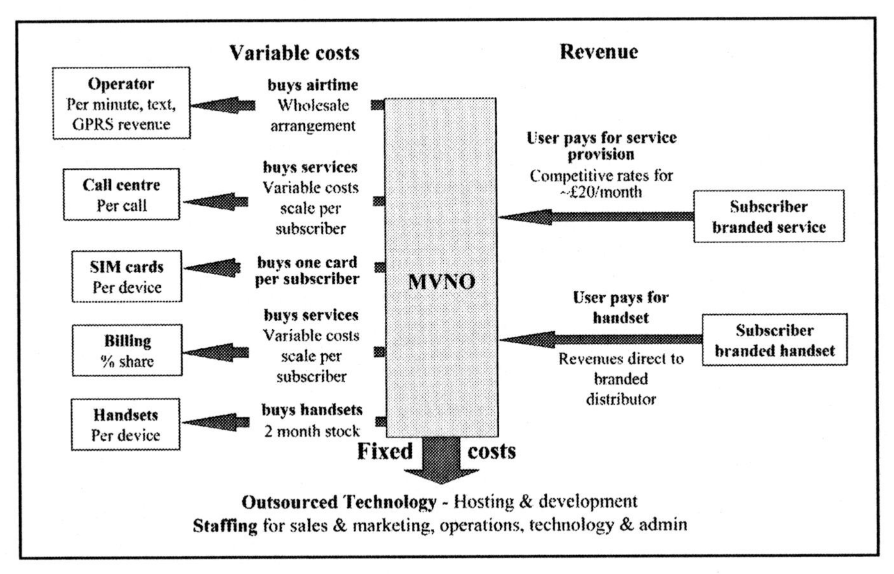

Figure 41. *The money flow around the business.*

The basis of the model is as follows:

- The MVNO buys:
- Airtime from an operator.
- Handsets from a handset supplier.
- SIM cards – typically from the operator but may be directly from a SIM card manufacturer such as GEMplus.
- Call centre and billing services.

- The MVNO sells:
- Contracts through its existing distribution channels.
- Handsets.
- SIM cards.

Although it will vary from deal to deal, a typical arrangement with a mobile operator would provide a model where every £1, $1 or €1 spent by the customer is split as follows:

50% pays for the airtime with the mobile network operator.
20% goes on subsidising the handset.
10% is spent on running the MVNO.
20% retained as profit.

How to start an MVNO

Starting an MVNO is not a trivial matter and should not be undertaken lightly, but with clear planning it is possible to minimise the amount of complexity that you will need to manage yourself. The very nature of an MVNO is to outsource the provision of the network to a mobile operator. Outsourcing should be second nature to everyone involved in this business. Ultimately, most of the functions of the virtual operator can be outsourced to one of four different suppliers:

- A mobile network operator – The mobile operator can normally provide SIM cards and handsets as well as airtime. Some mobile operators will offer technology, billing and call-centre services but you should think carefully before aligning too closely with a single supplier.

- A technology supplier – It is more important to get a partner with experience of MNO integration than to use a supplier you are familiar with. It would certainly be a mistake to use your usual technology supplier if they do not have experience in this area. Primary suppliers are IBM, LogicaCMG, and Ericsson, although you may be able to find an MVNE who can offer these services.

- Billing and collection – This could be part of the contract with the MNO, but it will probably be more cost effective to buy these services from an MVNE or a specialist business that provides similar services to virtual utilities. If you already have billing and collection services in-house, then use them.

- Call Centre – It is likely that your business will already have some call-centre arrangements and you should first consider adding the MVNO functions to this area. Alternatively you can outsource this to one of many businesses providing call-centre services. Some billing companies will also provide call-centre options and these can give good value for money.

Once you have set-up your main suppliers, you will need a small team of employees concentrating on developing your consumer proposition, establishing the channels to market, managing your suppliers and ensuring the overall profitable running of the venture.

The areas, which will most likely be new to your business, are:

- 24x7 operations of a business where consumers rely on the availability of your systems and will call (at your cost) to complain if they have any problems.

- Specific marketing for mobile phones.

- The problem of technical integration with mobile operators.

These are all areas where it is relatively straightforward to hire staff with the relevant experience: a good place to start is with the mobile operators themselves. Staff turnover among the operators is often high and therefore there is usually a pool of people available with the necessary skills.

Remember the main goal of establishing an MVNO is usually first to extend brand value and secondly to increase revenue. If you build a business with £10 million revenue but in the process knock 30% off the value of a billion-pound brand, then you have clearly failed. Basically, this means that you should not grow the business too quickly at the expense of quality of service, regardless of the level of income you may get from early customers. Growth in the mobile business is primarily driven by the acquisition of new customers. You will have a view of how many customers it is reasonable for your brand to achieve and you should plan on the basis that it might take as long as five years to approach the target subscriber level. Growth is therefore likely to be slow in the first two years of the business and you will need to inject sufficient capital to cover any potential losses made during this period. Figure 42 shows the typical timescales involved in setting up a new virtual operator.

Start Year 1	– Start building infrastructure and announce intention to launch MVNO
Mid Year 1	– Launch product with first customers
End Year 2	– Exceed one hundred thousand subscribers
End Year 3	– Record first annual profit
Early Year 4	– Exceed five hundred thousand subscribers
	– Reach break-even on original investment
Early Year 5	– One million subscribers
Early Year 6	– Sustainable profitability
	– Cash exceeds initial investment

Figure 42. *Timeline for the growth of a UK MVNO with a target of one million subscribers.*

The key performance metrics you should use are:

- New customers per month.
- Customer churn.
- Annual revenue per user (ARPU).
- Total number of customers.

Most operators have a churn rate of about 30% per annum, although some virtual operators with strong brands have managed to keep this as low as 15%. A key measure of the success of your MVNO is how your churn compares with regular operators: the lower the churn, the more value your brand is providing.

As the number of customers on your virtual network grows, you will need to expand the operational capabilities of your business. At the same time, you are going to be under pressure to continue developing the product to keep up to date with your competitors.

The most regular area of change is the cycle of new handsets. Most operators should be launching at least ten new handsets each year. At a minimum you will need to specify the configuration for each new device with your supplier, design the packaging, and test it in your own environment. Purchasing handsets directly from your network supplier is a great way to minimise your workload in this area, but you should not underestimate the effort needed just to stand still with the latest handsets.

Occasionally new network services are introduced. UMTS was the biggest area for 2005. In 2004 it was MMS, and undoubtedly 2006 will bring different challenges. New services require extensive customer communications, new handsets and often changes to billing. Value-added services such as ringtones involve a much greater level of change, but less impact on your every-day business.

Aligning closely with your host operator, using an MVNE, or hiring a key operations director from a network operator can mitigate all of these factors. However, you should not underestimate the operational complexity of starting an MVNO. Compared to other white-labelled products – such as credit cards and insurance – it is an order of magnitude more complex, but at the same time, significantly more profitable.

Conclusions

1. An MVNO strategy is good if you can reach a quarter of a million or more customers for your mobile services, ideally in a niche that mobile operators cannot reach easily themselves.
2. Launching an MVNO is a complex and difficult undertaking for any business and you should consider partnering with an MVNE to help you.
3. There are three generic strategies to choose from: brand, cost, or bundling.
4. Successfully executed, an MVNO can have a significant impact on the bottom line of even the largest of businesses.

Chapter 6
Everyone wants to text

" To read between the lines was easier than to follow the text."
– Henry James

Introduction

Text messaging, if you'll forgive the cliché, is the new rock and roll. In the 1950s and 60s, the baby boomer generation warmed to the new music as something that belonged exclusively to their generation: their parents didn't know much about it, those who did couldn't understand it, and ultimately many disapproved of it. SMS (Short Message Service) has had a similar effect in the schoolyards and colleges across the world. Young girls in Japan expect to receive four messages while asleep at night, or feel they aren't 'loved'; parents are ridiculed by the young for not knowing how to text; and a whole new set of language and terminology has emerged around it. It's no longer cool to play the guitar - Evry1Wnts2txt!

For the mobile operators, this provides significant additional revenue aside from voice traffic. For phone manufacturers, it encourages the development of handsets designed with text primarily in mind. For businesses outside the mobile industry, it provides a great opportunity to launch SMS-based services that can increase brand loyalty, encourage customer interaction and most importantly build revenues.

There are three key factors that have driven the growth in texting and its monetisation across the industry. These factors should be at the forefront of your mind in considering how your business can benefit from this area:

1. People will pay for texts – Although you will only get a small share of the money that the consumer pays, it is as effective a distribution model as retail magazines, and you can monetise electronic, time sensitive content, in a way that is completely impossible on the web or TV. People will frequently spend 50p (US$ or €0.75) on a premium text message and sometimes up to £1.50 for a service which is often very basic and cheap to run.

2. Barriers to entry are low – There are new businesses operating in the text area every month. All you need is a compelling idea, a little technical competence and the budget to advertise your services, and you are virtually guaranteed take-up.

3. It strongly appeals to the young – Texting is most prevalent among the under twenty fives, has some take-up between the twenty fives - forty fives but is virtually unused by the over forty fives. If your target audience does not fit this demographic then you are unlikely to make money out of texting.

Figure 43. *The division of UK SMS users by age in 2000. (Source: www.eurescom.de)*

The downside of texting is that the service you can offer is still only very basic. More sophisticated forms of messaging are available on mobile phones, but they still have only very niche usage amongst business users or employees[7]

[7] These people have their phone bills paid by their employer.

of mobile operators, although you can get comparable revenues to magazine subscriptions, you cannot sell SMS advertising in quite the same way.

Simply put, if you can design a basic service that you can market easily, then texting is the best way to start making money out of mobiles. Do it tomorrow... or even today!

The history of SMS

Initially considered an afterthought in the GSM specifications, SMS is a bidirectional service that allows you to send up to one hundred and sixty characters of text per message between mobile phones. It is different from voice in that the message is still sent even if the recipient's device is switched off or out of radio cover and is displayed to the user the next time the device is switched on. Messages are generally received within a few seconds of delivery and it is perfectly possible to substitute a two-way conversation with an exchange of text messages.

As SMS was only an afterthought, early user interfaces were very clunky and seemed clumsy to use when compared with speaking or e-mail. Sending text messages was not cheap in terms of cost per minute, but there were theories that it could become inexpensive.

An important thing to remember about text is that adults failed to predict that it would be children who sparked the massive growth in messaging. Quick to learn and with nimble fingers, they delighted in being able to do something their parents couldn't with the clumsy human-machine interface. Furthermore, while a voice call cost money, texting could be free since bright young mobile users quickly found a loophole; network operators were technically unable to bill pre-pay customers for text messages. It was only once the number of texts exploded that a more comprehensive charging system for texting was incorporated.

Despite the above mentioned text revolution, the interface for typing messages is quite clunky. The development of Q9 predictive-text-input software by Nokia may not produce as convenient a solution as using a QWERTY keyboard, but on the other hand, if the size of the mobile phone is to be kept compact, that is not really a choice. And Q9 certainly speeds up the texting process considerably. When you get beyond the Roman alphabet, things get more difficult - but that has not prevented the take-up of texting across the whole of the Far East.

The symbols employed in Chinese scripts may represent both sound and meaning. As a result, they include huge numbers (from hundreds to tens of thousands) of symbols. The Chinese also have a phonetic alphabet to make it easier to choose their characters or they can use stroke counts. In Japan they can and do text in four simultaneous alphabets and they use one of their phonetic alphabets to produce Japanese characters when needed. However, all require software-assisted input systems to generate the character choices and also require the user to learn the common characters to make inputting text a faster process.

The mobile phone operators were startled at the rapid growth in popularity of texting, especially as it happened without significant encouragement by them. Unsurprisingly when it did occur, promotion was quickly aimed at the younger generation. Texting rapidly became the preserve of school children using 2G networks to send their messages. However, despite the phenomenal growth in the number of texts sent, the price per message obstinately failed to reduce.

Business interest continues to grow with the significant adoption of mobile messaging. Interesting applications include vehicle-fleet management, and the automation of sales and servicing needs. Broadcasts of the latest news, public-transport problems or road-traffic congestion are another useful benefit. Financial information, event calendars and sports updates are also popular. In the public sector, success has already been achieved in congestion charging and parking and could be extended to health and other information alerts.

Some two hundred billion SMS text messages were being sent worldwide as early as 2001, a five-fold growth in just two years. By 2005, this figure had increased to one thousand billion, a further five-fold increase in four years. Predictions suggest that this number will reach some two thousand five hundred billion by 2010. Historically SMS has accounted for a high percentage of data ARPU, but inevitably this figure will fall, as 3G data services become increasingly popular.

On 3 December 1992, Neil Papworth of Sema Group sent the first commercial short message from a PC. It was sent to Richard Jarvis of Vodafone on the Vodafone GSM network in the UK. The design of SMS was originally part of GSM, but can now be used over most networks, including 3G ones. However, text message systems that do not use SMS include Japan's J-Phone's SkyMail and NTT DoCoMo's Short Mail. E-mail messaging from phones, as popularized by NTT DoCoMo's i-mode and the RIM BlackBerry, also typically use standard Internet mail protocols such as SMTP over TCP/IP, rather than SMS.

In 1998, the main UK network operators enabled the transmission of SMS messages between their networks. In April of that year, over five million text messages were sent. This number had exploded to over one billion by August 2001. Current estimates suggest the total may reach thirty two billion for the year 2005 in the UK alone, with peaks on Christmas Day, New Year's Day, Valentine's Day and GCSE and A level results days.

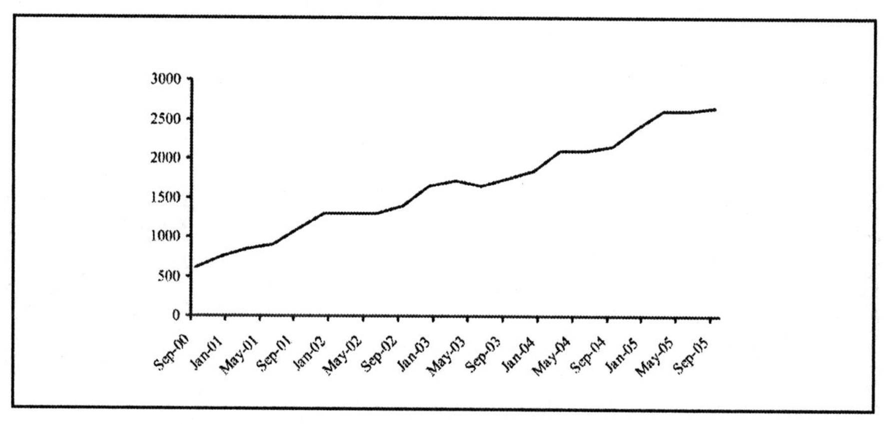

Figure 44. *The growth of SMS (millions of text messages) in the UK between September 2000 and September 2005. (Source: Mobile Data Association)*

The introduction of 3G phones offered alternatives to SMS, including instant messaging, e-mail and MMS (picture messaging launched in 2002) and the price of texting started to come down. While texting continued to grow worldwide, in the US it is still not as popular as elsewhere. Why? Probably the answer lies in part in the widespread availability of either unlimited or high monthly-voice minutes, and in the fashion of using P2T, that has proved so popular and gives instant connectivity.

As well as offering free browsing on operator portals, some networks offer free data usage for a specified time period. The fastest growth in texting was in the Far East and Europe, with the US lagging far behind, probably for the reasons mentioned above. Furthermore, the essential integration between competing providers and technologies for cross-network texting was slow to appear. Today the total figure is approaching three billion messages, worldwide, per day. However, in Japan, mobile text messaging is dominated by i-mode mail. SMS technology has only arrived relatively recently via DoCoMo's 'Foma' W-CDMA network.

Because of the limited length of message allowed and the awkward user interface on mobile phones, texters make widespread use of abbreviations. Numbers such as 4 replace words (for), vowels are skipped and spaces replaced with capitals e.g. ILoveYou. Much of this approach came from the contractions employed in Internet chatrooms to speed up typing. Texters who had to live without a full QWERTY keyboard rapidly extended its use. Similar solutions have been found for the many languages that do not use the Roman alphabet.

Despite these limitations, businesses are increasingly making use of texting. This not only appears to be the latest growth area, but software enables texts to be sent from PCs to one or more people at the same time. Even bus passengers can receive texts about when the next service is due by texting a designated bus-stop code. And this type of texting provides the opportunity to employ location-based services (see Chapter 8).

Consider the business proposition that texting offers mobile network operators. The incremental cost of running SMS to support around five billion messages a year is some £1 million-£2 million per year (US$ or €1.5-3 million). There is the potential for a five-hundred fold return on the investment, assuming a charge of 10p (in Europe or the US 15c) per text message or half that return if each message is half this price. However, there is still a concern among mobile operators that text messaging is cannibalising voice calls.

With the rise in the quantity of text messages, SMS spam has become the text equivalent of junk mail. It consists of text messages that may or may not be wanted by an individual mobile phone user. Three of Britain's mobile phone companies have adopted a code of practice, limiting the sending of unsolicited-text messages to mobile phones. In the UK at least, the Wireless Marketing Association (WMA) has issued a code stating that wireless marketing should only be sent to mobile phone users if they first give permission for it to be sent to them. Unfortunately the customer-complaints committee as yet has only limited powers.

One of the key factors driving spam is the low cost of sending a text message that users cannot avoid starting to read, than having to spend money on mailing out paper that will probably go straight into the bin unread. Despite this, brand names must avoid offending consumers by bombarding them with unwanted text messages.

In Japan in 2001-02, DoCoMo's systems were inundated by SMS spam, causing users' screens to freeze and even triggering the phone to dial an emergency number. In the US and Europe, on the other hand, texting is less popular than in Japan and furthermore, users are charged each time they send a message so that large volume spammers are put off by the cost; around 0.1 pence or cents per message.

Case study: French to regulate SMS costs

During 2004/05, the French regulatory authority ARCEP conducted a market analysis to determine whether a 'wholesale market for the termination of SMS messages' should be identified as a relevant market in terms of the application of the new EU regulatory framework for electronic communications, which had recently been integrated into French law. The survey was looking for evidence of the presence of high and non-transitory entry barriers or the absence of tendency towards effective competition.

An extensive questionnaire was developed covering retail- and wholesale-SMS markets to be completed in particular by mobile network operators. The questionnaire also contained questions about:

1. Market participants (end-users, aggregators and operators).

2. Prices and costs.

3. Competitive dynamics of SMS markets.

4. Presence or absence of significant market power.

The questionnaire recognised specific costs such as the expense of adequately equipped technical infrastructure at €3 million would impact on the 'overhead' cost placed on SMS.

The provisional conclusion was that single-network markets for wholesale SMS termination should be defined as a relevant market and that all three French mobile operators have significant market power for terminating SMS messages on their individual network. It proposed enforcing the following regulatory obligations on each mobile operator:

1. Commitment to provide access and interconnection.

2. Non-discrimination in the supply of wholesale SMS termination.

3. Obligation of transparency, including publication of the main wholesale charges.

4. Price control – an initial wholesale charge cap of €0.025 per SMS message termination.

5. Supporting cost-accounting and accounting-separation obligations.

The market analysis suggested that maintaining the wholesale charge cap at or above €0.01 per minute in the long run would preclude any risks of SMS spam. It also showed that the current wholesale charge between French mobile operators for SMS termination is €0.05336 and this has not changed since the launch of SMS interoperability in 1999.

Source: http://www.t-regs.com/

Despite these problems, annual messaging growth in Europe has ranged from 15-30%, driven by the escalating use among the young and improved uptake both by other age groups and business customers. As a result, SMS performance remains a crucial contributor to an operator's overall performance – both in terms of data as a percentage of total ARPU and also in terms of total-service revenues.

While the sales of GSM phones able to send and receive text messages has grown since their launch in the mid 1990s, the growth in the number of text messages lagged behind the introduction of these new phones by several years. Figure 45 suggests that the number of text messages sent by each user is also growing and is likely to continue to mushroom for the foreseeable future.

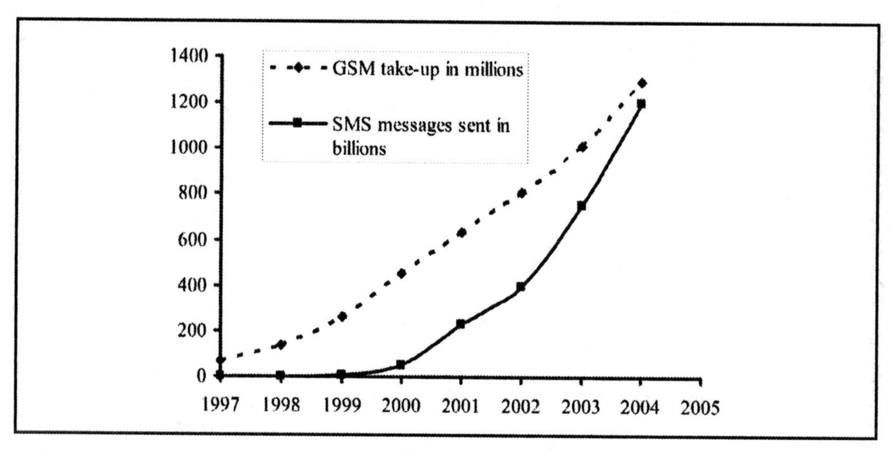

Figure 45. *Worldwide figures for GSM phone sales and number of text messages sent.*

Premium SMS

Premium SMS was first launched in the mid 1990s by Telecom Italia Mobile and is the natural extension of premium-phone numbers to SMS, providing third parties with the ability to receive text messages sent at a price premium by their customers. It has powered the explosion in mobile content services outside Japan[8].

[8] The content business in Japan was untypical in that it was driven by i-mode before the take-up of text, see Chapter 8.

Followed quickly by reverse-billed SMS, where the user pays to receive SMS, content providers rapidly had a large arsenal of mechanisms to bill for services on mobile phones. By 2004, this market had reached well over one trillion messages world-wide and spawned a new industry in its own right.

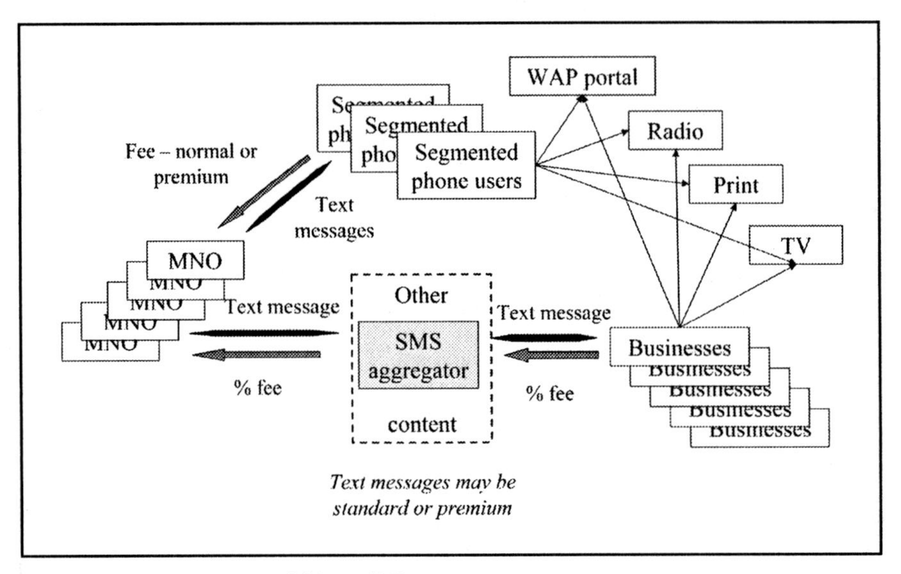

Figure 46. *A typical business model for an SMS aggregator.*

If you are intending to start a text-based service for your employees, customers or suppliers, you will need to get your services working on each mobile network that your target group will be using. Unless you are a mobile operator, you will typically want the service to work across all of the different networks in each country you are targeting, and mobile SMS aggregators have emerged to provide a single point of connection to businesses providing SMS services across different networks. The customers of SMS aggregators are not just media businesses or IT companies, as many mobile operators use aggregators to provide SMS interconnections to other operators where they have not negotiated an interconnect agreement.

Generally speaking, the market for SMS aggregation is crowded, with more than twenty operating in most European markets. This is a good thing for businesses, planning to use SMS in their go-to-market strategy, as it means that pricing and service are highly competitive. It does, however, mean that starting a new SMS aggregator would no longer be a wise business strategy and you should consider SMS aggregators only as potential suppliers for businesses looking to use their services.

Mobile aggregators will typically negotiate with the mobile operators to buy a number of short-codes that they can sell-on to their customers. These short-codes are typically between five and seven digits, and in many countries are regulated so that customers can easily see to what type of service they are subscribing. For example, in the Netherlands, all adult services begin with a particular set of numbers, as do all services that are subscription based.

Each number is typically associated with a price point and aggregators will either sell the short-code to a single customer or share it between several customers. For example, in the UK 63332 is a 10p shortcode that is owned by ITV, and 60060 is a 50p shortcode owned by the SMS aggregator Esendex. Esendex then sells keywords on to various businesses providing services on the shortcode. My business, TV Genius, rents the TV keyword on the 60060 shortcode.

Short code	Usage
50000 – 59999	Reserved for future short-code expansion.
60000 – 68999	Up to £1 per fixed fee SMS/call or for time-charged services up to £5.
69000 – 69999	Open-ended fixed-fee SMS/call or open-ended time-dependent adult-content services.
70000 – 79999	Reserved for future short-code expansion
80000 – 88999	Open-ended fixed-fee SMS/call or open-ended time-dependent services
89000 – 89999	Open-ended fixed-fee SMS/call or open-ended time-dependent adult-content services

Figure 47. *Basic UK allocation of SMS short codes.*

In deciding whether to use a shared-short code or buy your own, you should consider the following factors:

1. Owning your own short code means that instructions for your customer can be shorter and simpler. However, the price is correspondingly higher as you will need to buy the short code outright.

2. If you are using multiple-price points, you are going to have to use multiple-short codes. Remember the cost of buying them can rapidly increase.

3. Sharing a short code can cause problems. If users become unhappy with another service on that short code this may result in requesting the operator to stop providing all services on that code.

Overall, it really depends on how many services you are likely to launch. If you will have ten or more services, you are clearly better off buying your own short code and using that. If you're only running one service, then you are certainly better off using a shared short code unless you are certain you will generate significant revenues from it.

Operators	**Short codes**
3	62000 – 62999, 69200 – 69299, 69700 – 69799, 82000 – 82999, 89200 – 89299, 89700 – 89799
O_2	61000 – 61999, 69100 – 69199, 69600 – 69699, 81000 – 81999, 89100 – 89199, 89600 – 89699
Orange	60000 – 60999, 69000 – 69099, 69500 – 69599, 80000 – 80999, 89000 – 89099, 89500 – 89599
T-mobile	63000 – 63999, 69300 – 69399, 69800 – 69899, 83000 – 83999, 89300 – 89399, 89800 – 89899
Vodafone	64000 – 64999, 69400 – 69499, 69900 – 69999, 84000 – 84999, 89400 – 89499, 89900 – 89999
For example, the main UK operators share the following short codes to show a range of premium rates: 83080 – £0.25, 87080 – £0.50, 88080 – £1.00, 88600 – £1.50, 82772 – £3.00, 85080 – £5.00. They use unique keywords to indicate the network involved. A database shows the active codes, the available codes and the reserved ones.	

Figure 48. *Allocation of short codes to the main UK mobile network operators.*

As we have seen in the previous discussion about termination charges (see Chapter 4), it is usual for customers to pay only for services that they initiate. For an SMS service, this would typically mean that the user sends an SMS for which they are charged and then the user might receive an SMS in return that is paid for by the business that sends it.

For services such as one-off delivery of a ringtone or wallpaper, this mode is sufficient but it does not allow for efficient charging of subscription services such as football-goal alerts or share-price changes. For this reason, the concept of reverse billing has been developed. These two charging mechanisms are frequently referred to as MO and MT SMS:

1. MO stands for mobile originated and refers to any SMS sent from a mobile phone. MO charging therefore refers to a premium SMS sent from a user's phone for which they pay a premium.

2. MT stands for mobile terminated and refers to any SMS sent to a mobile phone. MT charging refers to a premium SMS sent to a user's phone for which they pay a premium.

As most SMS services operate a mixture of sending and receiving SMS, both options are a possibility. Typically you would use MO charging where a single text message must be sent to the user for a single billable item (e.g. buying a ringtone) and MT charging for subscription services like football-goal alerts.

Case study: Parker's Price Guide

Around a million people buy a used car in Britain each month – many of them roam the forecourts armed to haggle, clutching a copy of the famous Parker's Price Guide. The monthly guide contains the latest prices for over nine thousand models. But now Parker's has teamed up with Esendex to offer buyers personal, independent valuations of individual cars, by text message.

Customers, standing on the forecourt examining a vehicle, can text Parkers on 80806 with the word PRICE, the registration plate and optionally the mileage, for example: 'PRICE AJ51ELU 19756' Within seconds the Parker's price for that specific car comes back along with a security bonus – confirmation that the plate matches the model and make of car it is attached to.

With such an accurate price guide, buyers are better equipped to negotiate a fair deal. Customers pay £1.50 plus the normal network charge for the premium-rate service, which was developed by Esendex's Solutions Team. The facility hasn't been running long and has only been advertised on the Parker's website and in the Guide, but already three thousand people a month are using it and Parker's are delighted. Associate Publisher, Mark Wilson said: 'It's a valuable haggling tool when you're standing face to face with the dealer or private seller; I think it speaks for itself that already 5% of the people who buy our guide are using the SMS service. It's also been really easy working with Esendex, we've created a new revenue stream with very little effort or stress.'

Case study courtesy Esendex.com

An area of continuing contention amongst businesses that use premium SMS for billing is the revenue share that is taken by the operator. Many content providers complain that operator's commissions are excessive, compared with businesses like Mastercard or Visa, this is certainly so. Even when you subtract the cost of sending the SMS, the operator is making tens of percentage points in margin on the billing alone, compared to 3% for a typical merchant-services provider in other markets.

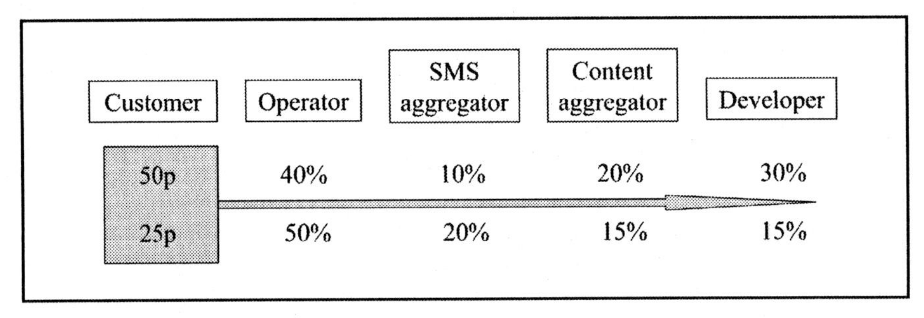

Figure 49. *Where the money goes in an SMS.*

Is this another case of mobile operators charging excessively for services? You must remember that running a mobile operation is a very capital-intensive operation. Shareholders will typically be looking for 40% EBITDA (earnings before interest, taxes, depreciation and amortisation) to justify a return on their investment. This makes it exceedingly difficult to run a low-cost billing operation and it is hard to see how the mobile operators will stay in this business over the long term.

As mobile services move towards an IP (Internet protocol) basis, it is certain that other billing providers such as PayPal will enter the market providing paid-for content on mobile phones. Indeed, synergies between PayPal and Skype were cited as one of the justifiers for eBay's acquisition of Skype in mid-2005.

The recent collapse of SimPay further highlighted this question. SimPay was a consortium founded in 2003 by three UK major mobile operators, Vodafone T-Mobile and Orange to promote and simplify payment for mobile commerce, by allowing customers to make purchases through their mobile operator-managed accounts. At the time it seemed to be a good idea. Other companies joined or expressed interest in joining the consortium; initially from Spain, then from Belgium, France and Germany. As a result of one of the founding members deciding not to launch the operation in mid-2005, this idea is unlikely to come to fruition. Many have speculated that the reason it stopped was that SimPay could not compete with the well-established credit-cards providers, Mastercard and Visa.

However, for all the problems, nothing has stopped the growth of payment using text messages. It is certain that for at least the next five years, SMS will prove a great way to sell mobile phone services.

SMS strategies

There are several ways of promoting SMS products; most phones have an SMS information menu built-in – within the industry this is normally referred to as the SIM toolkit menu. The operator will normally control and commission SMS services from third parties to go onto this menu. If you can get a deal with the operator, then you are likely to make a significant amount of money from it. However, be prepared to bow to every whim of the operator and support their frequently aggressive timescales.

The SIM toolkit menu accounts for about 30% of premium SMS traffic. Clearly there is a lot of money to be made by launching your own independent services. These are typically advertised in national newspapers or magazines and, as the cost of advertising frequently go into the tens of thousands of pounds, you need to have a very broad range of content to make this kind of advertising a compelling proposition.

If you can take out an advertisement for a couple of hundred potential purchases, you can reasonably expect a 3% take-up rate from the print advertising, so make sure you calculate the cost involved and divide it by the expected number of texts multiplied by the revenue per text. If the answer isn't a positive figure, you must ask yourself 'How can we make it positive?'

Txt 2 know

There are many reasons why customers are happy to pay for text alerts; generally related to something that is important to them. It may be a notable football match and they want to know whenever a goal is scored. It may be the value of a particular share on the stock market that they wish to sell when it reaches a pre-determined value. It may be that they wish to enter a final bid for an item on eBay just one minute before the auction closes and want a timely alert. They may simply wish to receive the latest news headlines or weather forecast, or their daily horoscope on the way to work.

Another example is the Google SMS service that you can use to search for a range of information from your mobile phone. Results come back in the form of a text message or SMS. You can send a message to their five-digit number followed by the word say 'McDonalds' to get the location of the nearest outlet. Google does not charge users to send a query or receive results. However, the rates that your mobile operator charges for sending and receiving a text message still apply. Google SMS presents the following instructions on its website:

Case study: Google SMS

Google SMS enables you to send queries as text messages from your mobile phone or device and easily get precise answers to your questions. No links. No web pages. Just text – and the information you're looking for:

- Get local business listings when you're on the road and want to find a place to eat.
- Obtain driving directions to get from point A to point B without having to ask for directions.

- Find film showtimes and cinema locations of films currently showing near you.
- Get quick answers to straightforward questions.
- Compare online product prices with ones you find on the high street.
- Look up dictionary definitions to expand your vocabulary or prove a point.
- Solve maths problems such as converting to metric units.

To use it

1. Enter your query as a text message.
2. Send the message to the short code 64664 (6GOOG on most phones).
3. Receive a text message (or messages) with your results, usually within a minute. Results may be labeled as '1/3', '2/3', etc.
4. To get Google SMS help info sent directly to your phone, send the word 'help' as a text message to 64664.

To learn more

- Find more information with our sample queries.
- Download our wallet-sized Tips Sheet (pdf).
- Do you have more questions? Look through our Frequently Asked Questions.
- Google SMS is available on all major carriers.

There are many opportunities for businesses to set up text-alerting systems either to improve their own efficiency or to improve their customer relations. The following example shows how finding a taxi in central London has been made much easier for customers and at the same time has almost certainly improved occupancy rates.

The London Taxi Point service uses SMS to enable mobile users to text and virtually instantly book the nearest vacant black cab. The service uses 'Taxi Points' that are physical signs carrying a unique four-digit code to identify their location within central London. People who want to use the service text the location code to London Taxi Point short code 83220. When the booking has been made, the cab driver can find the customer at an exact street corner or large building. Using GPS tracking, the service identifies the best-positioned cab and sends a confirmation SMS, followed by an alert when the taxi arrives. The service currently costs the user £1.

The following case study provides an interesting insight into another business use of text messages.

Case Study: Newcross Healthcare

Some time ago, nursing agency Newcross Healthcare realised that most of their four thousand nurses didn't have access to a computer and had to phone in to obtain necessary work information. Using the O_2 Business Text messaging service, work availability and pay-slip details are sent via text to nurses' mobile phones. The O_2 Business Text messaging service means that agency staff can be contacted without disruption to their work and confidential information can be delivered to a number of people in a fraction of the time taken via the telephone. The service also gives the agency greater flexibility and control over incoming-call activity and means that they can send birthday messages to their staff to promote increased staff satisfaction and loyalty.

Courtesy the Mobile Data Association Limited

Txt us

Clearly the entertainment business has latched onto the financial benefits of texting. The Live 8 charity concert did all its ticket allocations via SMS. Both 'Big Brother' and 'Who wants to be a millionaire' have raised revenue from getting viewers to send texts to the programmes. Sending text votes for the BBC sports personality of the year and the Eurovision Song contest have also proved popular.

Almost all UK TV channels now provide opportunities for viewer interaction by sending text messages as well as by phoning in. Comments on issues of the day and competitions are two good examples. Emap has used TV Channels to make money from SMS messaging. As well as using Freeview in the UK, it also has a number of pay-per-view channels. Viewers can interact with channels using text or voice to select music videos as well as to enter competitions. Even children's TV, such as CBBC, gives the chance to text feedback, and there is a guide to texting on their website.

It is the success TV, radio and advertising, aimed at encouraging SMS text messaging, that provides brilliant opportunities for audience interaction.

Big Brother 3 resulted in over ten-million texts sent to vote contestants off the show. Clearly, although mobile messaging has to compete with cable interface and digital set-top boxes, it is already popular with some ten percent of mobile users.

You can plainly see that the interaction of texting with TV has spawned a new generation of programmes such as quizzes, competitions and text-on-screen opportunities, usually with a background of musical videos or, late at night, with girls taking off their clothes, where texting involves flirting, dating or downloading pictures. All these programmes charge users on premium-call numbers or mobile numbers and that is how they make their money.

You will find a growing number of magazines that will also provide text messaging about breaking stories or updating existing stories. Hello magazine in the UK is a good example of this. Readers can sign up to receive daily jokes or cartoons. Betting and instant winning as well as lotteries provide mobile users with straightforward and easy-to-use methods of gambling. These may, however, be subject to national laws.

Responding by texting to advertisements, whether on radio, on TV, on advertising hoarding or in printed material, is an increasingly common way for potential customers to interact with suppliers about their products or services.

Often a competition is run for all those who text-in within a given period. The winner is given the product or service free of charge, but remember not to be over-optimistic about response rates: 3% is really the best you can achieve and a lot of competitions or free offers get less than 1% take-up.

Txt 4 stuff

Buying or downloading items seen either in advertising on posters, radio, TV, the cinema, in newspapers and magazines or on the Internet is a straightforward matter for any mobile phone user. They can download a ringtone or wallpaper and have the charge added to their mobile phone bill.

By far the most successful companies in this space make most of their income from the established ringtone market. However, they all offer a variety of content, some of it exclusive and developed in-house, and they are all making innovative use of mobile media and interactive SMS TV.

In Europe, there are four noteworthy companies in this sector. They have all built a market position in their home markets and then expanded into specifically targeted new countries and regions. All have sought financial backing, mainly from large and thriving Japanese or US corporations or sometimes from stock issues.

1. 123 Multimedia in France, is a company that was early to promote itself on TV. It is now under control of a Japanese company that also has a significant US presence and acts as its European arm. 123 Multimedia has its own TV channel to advertise content.

2. Buongiorno in Italy has to an extent lacked investment, despite a joint venture with a Japanese company, but has funded growth by issuing new share capital.

3. iTouch in the UK has now found a Japanese partner and has recently entered the mobile gambling field. It has used its 7777 short-code to help to develop its brand.

4. Finally there is Jamster in Germany, by far the largest of the four operations. Jamster is now American owned and has benefited from heavy investment. It is perhaps best known for its success with the Crazy Frog ringtone and undoubtedly has the strongest brand image. One clear factor to emerge is that scale of operation is the key driver in terms of profitability.

There have been some more innovative SMS applications that run using a GSM phone fitted with an appropriate SIM card to carry out e-commerce for each transaction. The SIM card interacts with both the network-being-used and the particular business site involved. It is suitable for tasks such as checking bank- and credit-card accounts, and for paying bills. In many ways a SIM toolkit can be likened to a smart card.

Barclaycard implemented the first application of the technology as long ago as 1997. A typical bill-paying transaction involves the phone user entering a PIN number, account details, amount payable and the date. Next this is automatically put into a text message sent to the vendor with an electronic signature. An acknowledgement text is then sent back by the vendor.

But the success of all of these applications has been far overshadowed by the ubiquitous ringtone.

Marketing and advertising using mobiles

Mobile phones open up a wide range of possibilities for marketing campaigns. For example, product information and special offers can be texted to existing and potential customers who have chosen to receive such messages. Strategically-placed poster and billboard advertisements, as well as ads placed in correctly consumer targeted newspapers and magazines, can include mobile numbers so that those interested can send a text message asking for further information:

- Special promotions can encourage customers to participate in competitions and win prizes by interacting with the brand by, for example, trying to win a free CD or DVD that has just been released.

- Mobile numbers can be printed on product packaging to encourage customers to text so as to give feedback or receive relevant information.

- Customers may be encouraged to forward messages about product offerings to their friends.
- Customer care is also an area where companies can use text messages to keep in contact with their customer base. Messages may remind customers about a forthcoming insurance-contract-renewal date or offer them product enhancements.

Undoubtedly advertising on mobiles will increase as 3G and later mobiles become more widely used but, if it follows the e-mail model, many people will consider it as something of a menace unless it is carefully targeted at the likely needs of mobile customers. Spam text messages are an ever-increasing problem. Those who use sharp practices or even resort to fraudulent ones do not help their image. An example of the former involves expensive premium numbers offering prizes and then putting callers who claim a prize on hold for as long as possible. When (and if) the case is referred to ICSTIS, the UK regulatory body for the premium rate telecommunications industry, fines only represent a small proportion of revenue to be earned. This should not be seen as a recommendation to set up such a business.

Mobile phone operators expect to sell airtime on their networks to advertisers, targeting their users with marketing and advertising campaigns. A company in the UK has already provided its customers with two free ads to download. You can successfully employ mobile marketing and advertising to:

- Increase public recognition, customer loyalty and the value of your brand by using mobile phones to offer new brand-based content.
- Increase sales by using a new avenue for promotion.
- Reduce the cost of marketing and advertising campaigns.
- Allow customers to opt in for mobile alerts for customised offers and personalised messages.
- Reduce the cost of sales by avoiding credit-card charges.

Successful campaigns that are examples of good brand marketing using mobile phones include the four-hundred-thousand PlayStation 2 users who responded to a request to send texts to their fathers on Father's Day and the campaign by US car manufacturer Pontiac. It asked camera-phone users to e-mail the company a picture of a Pontiac to enter a competition to win a free car. It obtained a quarter of a million responses.

Case study: Calvin Klein in Germany

The brief from the client

Figure 50. *Over 5.5 million cK one competition scratch cards were distributed through German cinema chains. (Courtesy MindMatics)*

The unisex fragrance cK one was revolutionary at the time of its launch in 1995. It was aimed at a stylish and brand-conscious target group (fourteen to twenty five year olds). However, the unisex theme in the cosmetics market had since been overtaken by other trends. In mid 2004, cK one was not even among the top twenty selling scents in Germany. In order to re-establish its former top-tier position, the client demanded an innovative media approach, reflecting the trend-setting image of cK one. The goal was to stimulate sales. Viral marketing mechanics were to be used to drive the campaign.

MindMatics' solution to the brief

MindMatics developed a cross-media mobile marketing campaign. The focal point of the campaign was a competition that was promoted via print ads, SMS, MMS-promotion teams, Internet and scratch cards. All entrants had to send in an SMS and received an SMS coupon that entitled them to a free sample of cK one. To generate a viral effect, customers were driven to a portal where they could send voice cards, SMS and scented photo-post cards to friends. Five million scratch n'sniff cards were given away to let the customer experience the fragrance.

The results
In September 2004, only four weeks into the campaign, cK one sales rose more than 500% compared to the previous year to become the third highest-selling fragrance in Germany, as reported by Europe's biggest perfume retail chain Douglas. The overall media strategy, the competition concept and portal services completely engaged the target audience. 30% of all competition entrants used the viral elements and thus extended the campaign reach far beyond expectations.

Courtesy: MindMatics

Evolution of texts

Will text messages survive in competition with e-mail, and vice versa? E-mail has a more user-friendly interface and people are very familiar with it from using it at work and at home. Furthermore, its informality when compared with a written letter has reduced the required number of words to convey a message. A QWERTY keyboard must be the first choice for entering the alphanumerics into any device. In favour of texting, on the other hand, are the further simplification and shorthand way of providing written messages, and the physical convenience of not having a relatively large keyboard. The convergence of mobile devices is examined in Chapter 9.

Beyond text – MMS

The multimedia messaging system MMS (text messaging with pictures, video or audio clips), developed for use with 3G phones, has yet to take off. This is despite starting to be deployed world-wide as long ago as 2002 across both GSM/GPRS and CDMA networks, and being operated by two hundred and fifty operators worldwide by the end of 2004. Unlike an SMS, an MMS is not transmitted via GSM but uses WAP protocol (GPRS or UMTS). This means that to send and receive MMS messages involves the network operator in providing additional infrastructure. There are still some problems with other-user compatibility when pictures are included and sending messages to multiple addresses is a complex operation.

However as the number of camera phones rapidly grows, it is likely that more and more texters will be drawn from SMS to MMS, though the clunky interface is likely to prove something of a deterrent. Furthermore, mobile operators still fear that too many MMS messages could swamp their networks and thus place limits on number of these messages. Equally apparent is the fact that MMS offers great opportunities for media companies. The monthly number of MMS messages sent in the UK was only around thirty million at the end of 2004, perhaps one percent of the SMS total, and an estimated sixteen billion worldwide. Even if MMS really takes off, it is unlikely to kill off SMS, which still represents a cheaper and easier method of sending messages. Furthermore, MMS is still not used to make mobile payments.

Case Study: BBC News and MMS

BBC News have for some time been asking people to send in photographs of major news incidents by MMS.

On the 7th July 2005, they received a thousand pictures of the terrorist attack on the London tube and buses. In many cases, these provided the only footage of the news at it happened, and formed a key part of the coverage on both the Internet and TV.

Conclusions

1. The revenue-earning capabilities of SMS are well proven, as are its benefits in marketing and advertising.
2. You can probably make more money in the short term out of SMS than you can from any other strategy, unless you are planning to start an MVNO (see Chapter 5).
3. With many incumbents already, it is important to find an innovative idea and to be clear of your target market.
4. There are still serious doubts whether MMS will generate significant revenues.

Chapter 7
Handset mine

"Looking good and dressing well is a necessity. Having a purpose in life is not." – **Oscar Wilde**

A key factor in the success of mobile phones is not so much that they are mobile as that they are personal. The driver for use at home was the ability to store your personal address book in your phone and it did not take long for a phone to become a fashion statement about who you are. Beyond the simple choice of which phone to buy, the manufacturers soon put in configurable ringtones, wallpapers and changeable faceplates.

The growing trend in personalisation was missed by the operators, who concentrated like so many companies in engineering fields do, on the technology and the requirements for the next generation. In doing so, they took their eyes off the opportunities presented by the existing products. As such, the scene was set for new entrants to start selling ringtones and wallpapers to customers. Although mostly based around pop stars and celebrities, problems of licensing and the quality of possible products made the record companies cautious about entering this market. Unsurprisingly a range of new businesses took off, recording home-grown versions of popular songs and scanning in celebrity pictures from magazines, to be sold by SMS.

Most of these businesses launched around year 2000. At the same time the music industry was reeling from Napster, and companies like Handy.de quickly became the most popular destinations for mobile users in Germany with over five million people registering to use mobile-entertainment services such as ringtones, graphics, downloadable games and other mobile fun.

It didn't take long for mobile operators to recognise what they were missing and to climb on the band-wagon, launching services like Vodafone Live!, O_2 Active and T-Mobile's t-zones. All were aimed at providing customers with appropriate content to personalise their mobile phones. Now, five years after the initial spurt, the music publishers are finally entering the market with ring tones sold alongside CDs and singles.

Although there have been some sales of non-personalisation content (see Chapters 8 and 9 for more details), the vast majority of mobile revenues relate to content that enables people to customise their phones. The sector grew quickly and was worth over £1 billion in 2004. It is still both growing and developing, with many new products such as ringback tones and games driving growth, but to date the most successful products are still ringtones and wallpapers.

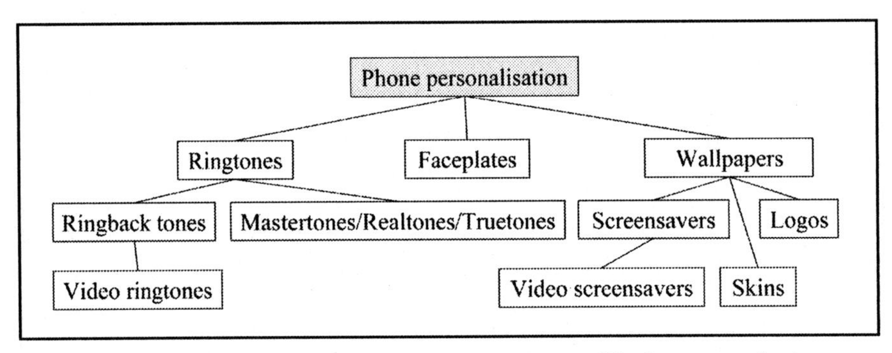

Figure 51. *There are many different components that go to make up mobile phone personalisation.*

Personalising your phone

If you have a compelling consumer brand that you can use to produce content to personalise mobile phones, then this is probably an area where you can make money. Broadly speaking there are three different categories of personalisation content:

1. Ringtones.
2. Wallpapers.
3. Faceplates.

What all of these items have in common is that their primary functions are to make the phone a more attractive device to the user and to communicate something about the user's preferences to other people in their environment.

Ringtones

Original ringtones were purely monophonic and played a single melodic sound in a standard beep for each mobile phone. These are still a major revenue earner, but most modern phones have replaced monophonic ringtones with polyphonic ringtones which provide synthesised sounds with multiple instruments and even drums.

Polyphonic ringtones are significantly cheaper to produce than monophonic ringtones because they use the standard MIDI format across all different phones, whereas monophonics have different formats for different handsets. Combine the lower cost of production with the perceived increase in value from a user's perspective and you have a significantly improved business proposition.

The latest generation of ringtones and actual clips of music songs or instrumentals, frequently in MP3 format, and are referred to as Realtones, Truetones or Mastertones, across the industry. Because they use real clips of songs, they cost more in terms of licence fees. This cost is charged on to the users, who will require more sophisticated mobiles for the ringtones to work correctly.

Other ways that are currently being trialled to increase revenues from ringtones include video ringtones and personalised ringback-tones, where a custom tune is played back to the caller, rather than the usual ring-ring. It is certain that there is a long way to go before the market for ringtones is even close to saturated.

Case study: The Crazy Frog ringtone

The Crazy Frog ringtone, which dominated the airwaves during the summer of 2005, originated in the late nineties when Daniel Malmedahl, a young Swede, recorded himself attempting to imitate the sound of a motorbike and posted it on a website as an MP3 realtone. The idea was developed on the Internet, on Swedish TV and with 3D animation. The animation and sound were licensed for use as a mobile phone ringtone and marketed by Jamster. It achieved exceptional sales, over £40 million in 2005 from the ringtone with virtually zero production or distribution costs, making it the most commercially successful ringtone ever.

The sound of the ringtone proved a great irritation to many adults and its originator even apologised for this in a BBC interview. It also generated a number of complaints to the Advertising Standards Authority, subsequently not upheld, the first on the basis that the Crazy Frog appeared to have genitalia, making it inappropriate viewing for children. The second objection was based on the regularity of the appearance of its ads on TV; often more than once per commercial break and over seventy thousand times in a single month – an expensive campaign.

Perhaps more unfortunate was that written in the small print was the requirement not just to pay a single small fee for the download, but that Crazy Frog was only retainable by paying the same fee as part of a regular subscription, making its use an expensive proposition. This complaint was upheld.

Despite this, the Crazy Frog as the most popular ringtone download has spawned a number of music singles, including a number one hit. It has also resulted in a host of other related opportunities, ranging from video games to i-tunes downloads.

Wallpapers

Before the advent of camera phones, wallpapers were a 'must have' item of personalisation, if only to differentiate phones from others in the same household or office. Wallpapers' sales, although still greater than any other category of content except ringtones, are now steadily declining, with most people using the built-in camera to create their own wallpapers.

The first wallpapers were called operator logos, and were original monochrome images used to display the mobile operator's logo rather than their name in plain text. They changed when connecting to a foreign network overseas and provided only a very basic form of personalisation. However, it was enough to make sure you did not pick up the wrong phone.

As phones developed, the clear requirement for colour screens and better graphics has resulted in wallpapers that cover the whole screen, in a similar way to those provided for PCs. As with PCs, people enjoy the individualism that a quality wallpaper provides, often buying pictures of their favourite celebrities, and frequently changing them.

As built-in cameras have made many types of wallpaper obsolete, developers have extended the metaphor to screensavers that can be simple – frequently amusing – animations, or more expensive video clips from pop videos or films. They can be popular with the younger mobile users but they also place greater demands on battery consumption and processing power. Despite this their use is set to increase.

The ultimate in phone personalisation is a themed package based around a single theme or artist such as Michael Jackson or Kylie Minogue. A package typically comprises ringtones, screensavers and wallpapers, together with customised menu items and icons on the phone. The arrival of new icons on the pop music scene provides continuing opportunities for new skins, but to date these have not had the penetration of the individual items. This may be partly due to the total cost of the package but also that the personalised menu items and icons only function fully on high-end phones and require significantly more time and effort to develop.

Faceplates

Faceplates are physical add-ons for your mobile phone and must be customised to fit a particular handset. They do, however allow customers to make their phone look physically different and may be used to match preferred colour themes. They are particularly popular with female users and fans of football clubs who are partisan in their apparel.

Faceplates are a large market in their own right, but because of their low-tech nature are sold through market stalls and other more traditional distribution channels than most of the topics covered in this book.

Personalisation	Device demands	Ease of network operation	Complexity of licensing	Uptake & price
Ringtones	Low	Low	Simple	High usage/ revenue market
Logos, wallpapers & screensavers	Low	Low	Simple	Growing as colour mobiles numbers increase
Realtones & Truetones	Medium	Low	Complex	Low volume, premium price
Themes	Medium	Low	Relatively complex	Low volume
Video ringtones & screensavers	Medium	Medium	Complex	Emerging

Figure 52. *The market for mobile personalisation in Europe.*

Case Study: McDonald's Finding Nemo

In 2003, McDonald's and 12snap, a Munich based provider for mobile marketing and entertainment solutions, implemented one of the most successful mobile marketing campaigns ever. The campaign target was to drive sales and increase brand preference by creating a one-of-a-kind mobile marketing on-pack campaign in Germany, Austria and Italy based on the Disney/Pixar movie Finding Nemo, targeted at the youth market. Partners of the promotion were Nokia and Vodafone.

Figure 53. *The drinking cup, the code revealed, the advertising banner and a Finding Nemo figure on a Nokia phone. (Courtesy 12snap)*

Twenty five million drinking cups were provided with unique eight-digit codes. Consumers sent in the code via text messages. As a reward they received one of seventy mobile specials based on Finding Nemo such as ringtones, a Java game, wallpapers, personalised wallpapers, postcards and animations. The on-pack campaign was communicated by a consistent appearance in the media. TV, print and radio commercials supported the announcement of the promotion. At the point of sale, traymats and posters backed the announcement of the promotion.

The result was phenomenal. The response rate reached an incredible 18% (four and a half million participants). The original target was exceeded three and a half times. All objectives were fulfilled.

The key success factor was that the mobile campaign was not merely a side promotion but rather the main focus of the campaign and therefore perfectly integrated in the media mix. It was clearly visible to the consumer that it was a mobile promotion. The mobile phone was always in the centre of the promotion.

The explanation of the SMS&Win mechanics on the traymat was easy to understand, allowing customers quickly to grasp the concept. This led to direct response from the point of sale which boosted response rates. By integrating a pull-mechanism, McDonald's also had a perfect response tool to create a dialogue with its guests.

The content of the promotion was extremely innovative, unique, entertaining and attractive as it was based on the successful movie Finding Nemo. In 2003, the average consumer price for mobile content was very high (around €2-5). In the Finding Nemo campaign consumers received the content free by buying a drink at McDonald's. This was a great incentive for the consumer to participate and provided them with additional fun in the restaurants – a reason to come back again for further participation.

Courtesy: 12snap

Operator portals and content aggregators

By their nature, most personalised services reside on the mobile phone and may either be pre-installed on the phone at the time of purchase, purchased as a hardware plug-in (e.g. flash memory or faceplates), or downloaded onto the device by the end-user.

Mobile phone sellers work with content providers to pre-install personalisation in the handset, but more often than not, content is provided via SMS or WAP portals with the charge being put straight onto the phone bill. While SMS began as the dominant medium, and the glossy magazines are still full of advertisements for ringtones, WAP portals have become much more widely used since the introduction of colour phones. An increase in downloads of personalisation content from these portals is likely.

Most mobile operators run their own portal that is easily accessible from a single keypress on the phone. On these portals you will find a range of different ringtones, logos, screensavers and wallpapers. Some of these are available directly from the operator; others are obtainable from third parties using SMS text messages. The choice is enormous and these portals are subsequently expensive to maintain. Operators have to make a choice about whether to be serious players in the content game themselves, or to leave this to the independent portal providers.

While most operators continue to grow their personalisation portfolio, they still tend to focus on the technological rather than user requirements of the proposition. They are under pressure to be seen as proactive in the music arena, but are aware of the difficulties in offering music services. They launch ringback tones but find that users do not take them up as extensively as conventional ringtones. They secure very-expensive deals with record labels and films studios but find it difficult to make a return from this part of their businesses.

Meanwhile the independent portal operators like Jamster are focused solely on building revenues from the content business and they are very successful at it. Despite the close integration between the operator's portals and the handsets, more than two thirds of content revenues go to the independents.

All of these independent portals are run as content aggregators: they put together portfolios of ringtones, games and fun, screen pictures, music and videos. Different aggregators target particular market segments, and many also run some sections of operators' portals.

As with operator portals, most billing is provided by premium SMS (See Chapter 6 for more details). Products are usually priced in one of two ways;

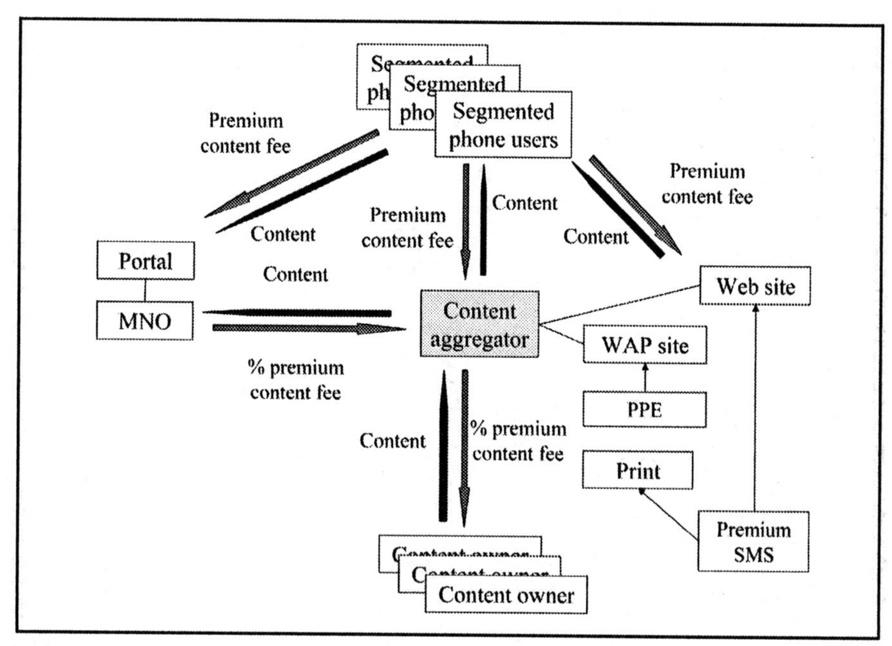

Figure 54. *A typical business model for a content aggregator.*

either there is a one-off charge for a download or there is a monthly charge for as long as the user continues to enjoy the benefit of the content. Remembering that the operator takes the lion's share of low-cost transactions, it is important to decide how many customer-downloads at what price-per-download or per-month will be necessary to support your cost of operations and provide a reasonable level of profit. Unfortunately users tend to prefer a one-off charge to a regular monthly one, however they may be put off if the former is too large and because of the large operator share it may be necessary to inflate the price of one-off transactions beyond what you may feel is reasonable. But do not fear. Mobile customers are famously price insensitive. After all, it's only a text, how much can it cost?[9]

The traditional way of marketing mobile content is on the basis of direct advertising, using TV, radio, newspapers and magazines. All of these media concentrate on particular market sectors and this is an advantage for content aggregators, since they can target their potential customers more accurately.

9. Texts of up to £1.50 are common in the UK, with higher values possible.

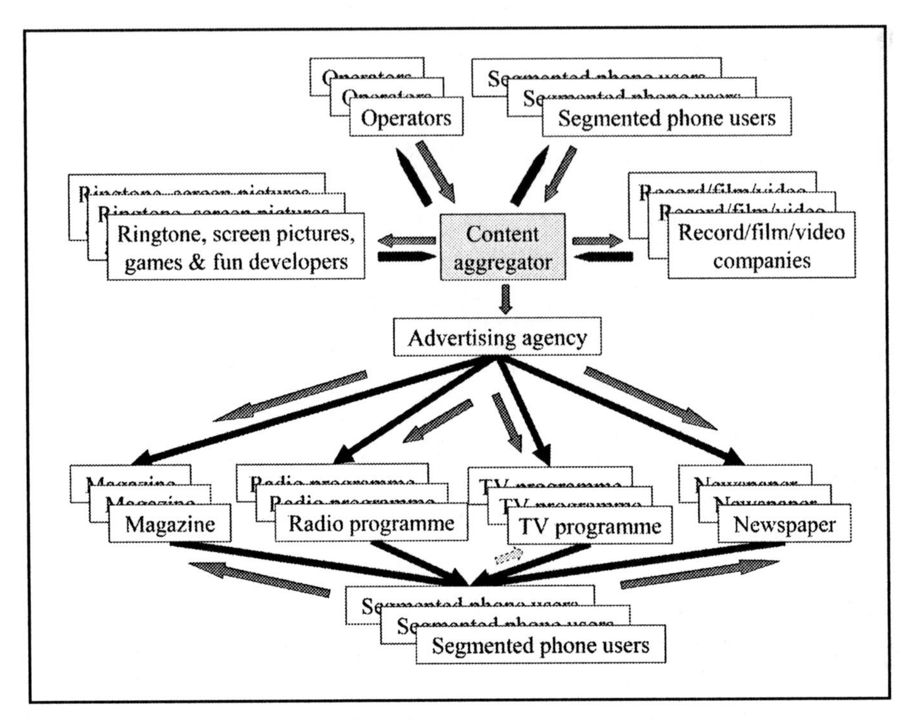

Figure 55. *A typical model of a direct-advertising content aggregator.*

However, most publications nowadays are reluctant to take too many advertisements for mobile services because the adverts themselves are often unattractive in design terms. The advertisements are very busy and densely packed with hundreds of different ringtones and wallpapers for sale. The Editor of one of the UK's biggest-selling magazines told me he had to limit the number of these ads, otherwise they would dominate the look of his magazine. All of this means that advertising costs remain high, and aggregators will need a broad range of content in order to make the business profitable. The very nature of personalisation is that people do not want to buy what everyone else has and you therefore need to be able to offer sufficient range so that people feel they are getting something unique.

Selling your content

If you are looking to sell your content to personalise mobile phones, you have three different routes to market:

1. Direct sales to your customers - The most profitable if you have a cost-effective route to attracting customers.
2. Through-deals with mobile operators - Only really valuable if you can persuade an operator of the benefit of making your content available exclusively on their network.
3. Through-deals with content aggregators - The de-facto standard approach.

Case Study: Warner Brothers' deal with Verizon Wireless

Verizon Wireless is obtaining mobile content from Warner Bros (WB). Scenes and trailers from the WB network shows 'Everwood,' 'Gilmore Girls,' 'Smallville' and 'Supernatural' will be available through its VCast service. Both classic and original Looney Tunes and Hanna Barbera cartoon shorts will also be offered including characters such as Bugs Bunny, Tweety, the Flintstones, the Jetsons and Scooby-Doo. The VCast service will also feature trailers from upcoming motion picture releases from WB, Warner Independent Pictures and Warner Home Video.

VCast is priced at $15 a month in addition to a Verizon calling plan. Application download fees apply for 3D games and premium video, but there are no airtime charges for downloading or streaming VCast content.

The typical route for direct sales is through cross-promotion from other products. You may sell products to customers that can have mobile services on the packaging or in the product themselves, or you may have a low-cost way to distribute leaflets or similar to customers. In general, a large Internet presence does *not* correspond to large mobile sales. You really need to be effective in the physical world if you are going to promote mobile content through cross-promotion.

If these options are not open to you and your content is highly valued by a large number of consumers, then you may be able to get a deal with a mobile operator. Typically these work on a fixed licence fee that the operator pays up front in return for the right to sell content only to its own customers. You could combine this with a revenue share on any sales, but with an exclusive deal it is more common to look for a licence fee up front.

Selling through content aggregators, including non-exclusive operator deals, is normally relatively straightforward once you have found the right people to deal with. Buyers within mobile-content businesses are famously busy and it is hard to get hold of them. Most aggregators are secretive about their business-contact details and *just getting* a contact phone number can be difficult. Once you are in and talking to the right people, you will normally be offered a standard contract which is non-negotiable and offers a very low revenue share for your content.

If you agree to the deal, the onus is then on the content aggregator to promote your content. A deal with an aggregator is the equivalent of a record contract. It has a lot of promise but is only really a beginning. For every Beatles or Rolling Stones there are thousands of other acts with record contracts that never achieve anything. The same goes for personalisation contracts: nearly ten years and there has only been one Crazy Frog. Of course, if your content is the next Crazy Frog, then you really have struck gold!

Setting up as a content aggregator

If you do not own the rights to any content yourself, you could still consider setting up as a content aggregator. Although the market is already crowded, most incumbents are targeting the same sector: primarily youth. There are clear opportunities to provide mobile personalisation to other different sectors. To enter this business, you will need to consider the following five factors:

- Market segment – There are numerous ways of segmenting the market, for example by age, by ethnicity, by gender, by income or by county. If you are looking to target, say, the over fifties, your ringtone selection is unlikely to include Crazy Frog and 50 Cent so you will need to source content effectively. Remember too that you cannot offer exactly the same content worldwide, as language and cultural differences come into play.

- Content – As a new entrant, it is the content that will differentiate you from others in the market. This is going to involve setting up contracts with a range of content developers, including dealing with IPR, and a brand can be a key attraction to potential customers. Remember that your content must match your target market.

- Distribution – Once you have determined what you are going to sell and secured the rights, you must have a clear way to gain access to your market. To date, content has responded well through advertising in newspapers and magazines, with significantly lower sales from the web, although direct marketing to a database (perhaps combined with an MVNO approach – see Chapter 5), could yield a high return.

- Business model – You must have a carefully constructed business plan. Avoid the trap of launching a business and trying to capture customers on the basis of low prices alone. If anything, you should look to sell content at a premium in the early stages to early adopters within your market segment. If you cannot justify a price premium, then you are probably competing in an already over crowded market and are unlikely to achieve long-term profitability. Know the market price for your offerings and aim to get those prices in the majority of your sales. If you are offering premium content, then you should expect to obtain premium prices, as you should if you segment early adopters who always seem anxious to buy the latest and coolest, regardless of the higher cost.

The right payment procedure can be critical. In the mobile sector, premium texting has always been preferable to the use of credit cards. This may not always hold true, particularly in less mobile-savvy segments, but if you do bill by SMS, make sure that you check and adhere to local and international regulations and note that these are continually changing. Remember that laws differ considerably from country to country. What may be legal in the UK may be illegal elsewhere. Even restrictions placed by the operators may have an impact on the way you construct your business model.

- Technology – The basic systems required to provide personalisation content are not overly complex and there are many off-the-shelf products that can be bought, but it is still usual for content aggregators to develop these systems in-house. Indeed many profitable content aggregators are exceedingly small businesses which run a small amount of technology, license (or occasionally fail to license) content, run an outsourced customer-support centre, and take out advertisements.

The biggest challenge you will face as a content aggregator is long-term growth: in the long term, the market for personalisation content is limited and as phones become more PC-like there is even a chance that ringtones may be downloaded free over the Internet. Even if it continues to be paid-for, the major media businesses will look for their share. Coldplay's hit album of X&Y, released in 2005, came with a sticker on the front cover listing the shortcodes to be used to purchase ringtones of each song, and ultimately it is likely that the market will fragment along these lines.

Conclusions

1. The world personalisation business is huge - worth some £5 billion at the end of 2005. Most barriers confronting growth in the provision of phone personalisation are technical rather than cultural and are already being addressed.

2. The penetration of phones able to play real music is growing with video capability certain to follow. Digital-rights management is still an issue and both record and movie suppliers are jockeying for position.

3. There is a question of whether MP3 players and their video equivalents will remain as self-contained devices or disappear as these facilities are merged into new generations of mobile phones. This is discussed in Chapter 9.

4. If you are a media company with an idea that can provide unique content, sell it through a content aggregator. Setting up as yet another content aggregator is not recommended.

5. Do not consider setting up any site that could be considered to support either free downloads of copyright material or users illegally sharing copyrighted content on their networks. You are breaking the law and likely to face huge settlements as the recent Grokster case in the US has highlighted.

Chapter 8
The Internet on your mobile

> *"The Internet is like alcohol in some sense. It accentuates what you would do anyway. If you want to be a loner, you can be more alone. If you want to connect, it makes it easier to connect."* – **Esther Dyson**

It is amazing to realise how recently the Internet and the World Wide Web arrived. Yet, it is only in the last decade that they have become a widely available and practical proposition. The Mosaic web browser was first offered in 1994 and Microsoft Internet Explorer is only just ten years old.

The provision of online services in the US dates back to 1969 with the launch of Compuserve (later purchased by AOL). The first DOS-based version of AOL was launched in 1991 and a Windows version appeared two years later. Microsoft's failure to purchase AOL in 1992 led to the subsequent launch of MSN in 1995, by which time AOL had become recognisable as what is today called an Internet Service Provider (ISP).

By 1996 AOL had five million customers built on a reputation for quality of service and a large array of exclusive content for its customers. All of the major ISPs were in the process of signing up deals with media businesses to promote their portal products based on content that could only be consumed if you used that ISP.

However, half way through 1996 AOL announced its 20/20 plan in the US – twenty minutes for twenty dollars – reacting to the lowering of rates by competitors, and in October MSN announced a flat rate charge of $19.59 for unlimited Internet access. Six weeks later AOL was forced to match this price and suddenly the market had moved from ISPs' content to cheapest ISP price. By the end of the decade, AOL was still in the number-one slot, but being pursued by five other providers for second place. Many of the smaller and mid-sized ISPs were offering free-of-charge services, using revenues from the telephone-line provider and/or advertising income with little, if any, exclusive content.

Many governments, particularly in the industrialised nations, started to express interest in the percentage of households with Internet access, recognising the many benefits that such access would bring to their populations. Although few countries actively subsidised the Internet roll-out, most have indirectly invested millions through the development of Internet-based government services, such as online tax returns.

Broadband connections to ISPs emerged around the Millennium and not surprisingly triggered a price increase. However, the 'always on' capability, allied to the faster download speed rapidly overcame market resistance to the cost. In addition, Internet search engines like Google simplified the locating of information on the Internet, and opened it up to a whole new generation of less technically-savvy users.

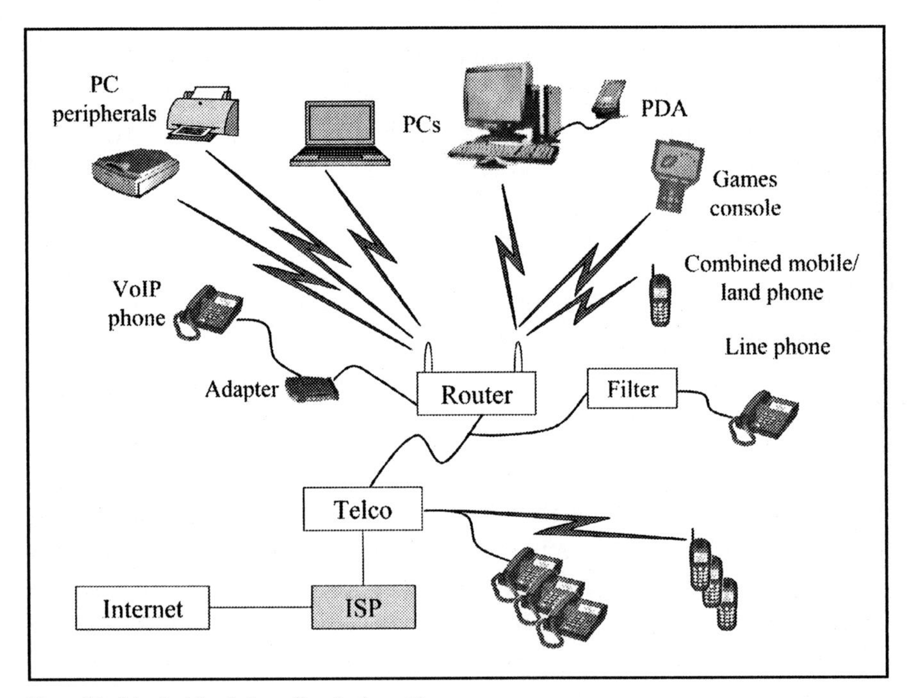

Figure 56. *A typical family broadband set up at home.*

In the UK, Freeserve was set up as an ISP by the high-street electrical chain Dixons. Its attraction to users was that, as its name implied, it provided free dial-up Internet access. As a result, it experienced phenomenal growth as an ISP, and shortly after its launch in mid 1998, it had achieved half-a-million customers and by the spring of the next year this number had grown to one and a half million. The success of the no-charge Internet access approach resulted in Freeserve providing its services to one-in-three UK Internet users within its first nine months.

In 2001, Wanadoo, a wholly owned subsidiary of France Télécom, acquired the company. It subsequently broke its links with Dixons and was re-branded as Wanadoo, currently the largest ISP in Europe with more than ten million subscribers, mainly in the UK, France, Holland, Spain and a number of Francophone countries. It is interesting to note that is has now broken away from the free dial-up access and although it received the majority of its revenues from access fees, its website still offers an extensive content catalogue including news, entertainment, sport, travel, and games. France Télécom has announced that all Wanadoo's services will be re-branded to Orange, its MNO, in 2006 to simplify its branding, and we can thus see clear signs of convergence between the ISP and mobile operator sectors.

So how do ISPs make their money? Initially, fees were subscription based using a credit card and a local dial-up number. Outside the US, where local calls are not free, this quickly moved to a free service where the ISPs obtained a percentage of the cost of dial-up calls. In addition they gained revenues from companies that wished to advertise or have a hyperlink directing users to another website. Advertising models on the web have matured and it is now common for ISPs to make 10-15% of their total revenues from advertising on their websites.

As broadband took over from dial up, the model changed somewhat with all users paying a flat-rate subscription, part of which was paid to the telco for access to the line, but was all billed by the ISP. As competition in the broadband market increased and the speed of broadband was raised, some pricing models changed .A range of different speeds at different prices were offered with a cap on the monthly amount of downloads beyond which a usage fee would be charged, but the basic model is unchanged as shown in Figure 57.

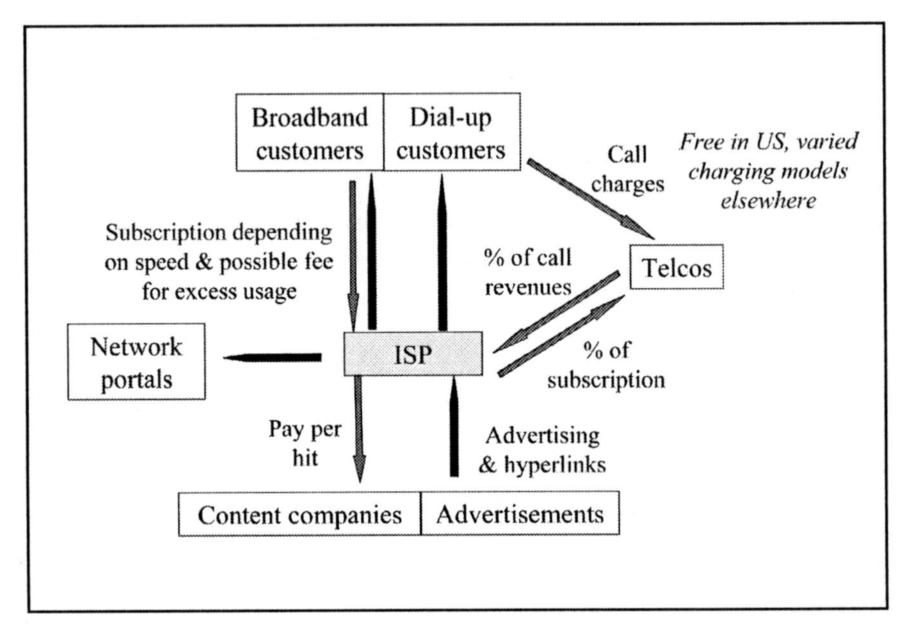

Figure 57. *A typical ISP business model for dial-up and broadband customers.*

The dot.com crash

From 1996 onwards, it was clear to most large companies that a web presence would be important. Once people became more familiar with using the Internet for gathering information and fast group communications, the potential attractions of e-commerce became apparent. These concepts fascinated many young entrepreneurs who recognised that new business models could be based on these possibilities and wanted to profit from using them.

The arrival of an inexpensive way of instantly communicating with millions of customers worldwide suggested new ways of advertising and selling by mail-order. The Internet would be the latest low-cost vehicle for bringing together sellers and their potential customers. Young graduates in the industrial nations developed previously unimaginable business models and rapidly succeeded in selling their ideas to venture capitalists.

In the late 1990s, dot.com companies were market favourites, with both speculators and investors pouring cash into new start-ups that could exploit the increasingly pervasive Internet. One result was that many youthful professionals from conventional companies were attracted to work for these start-ups in California and later the UK Thames Valley. Many of these young enthusiasts lacked any experience of the Internet. The public became bemused by the opportunities the Internet presented and financiers were prepared to fund new dot.coms almost regardless of whether their ideas were supported by a practical business plan. IPOs (initial public offerings) or the first public sale of stock by in this case Internet companies, appeared with incredible rapidity, exciting investors in the both the US and the UK. The shares in some 25% of them doubled on the first day, even though the vast majority had as yet made no profit. Venture capitalists, seeing the fast rise in valuation of dot. coms and moving with more speed and less caution than usual, spread their risks by financing many dot.coms and letting the market determine the successes. The CEOs of the businesses were left with pressure from speculators to spend their millions of dollars or pounds of investors' money to 'become a Microsoft.' It was clear that this would result in many more losers than winners.

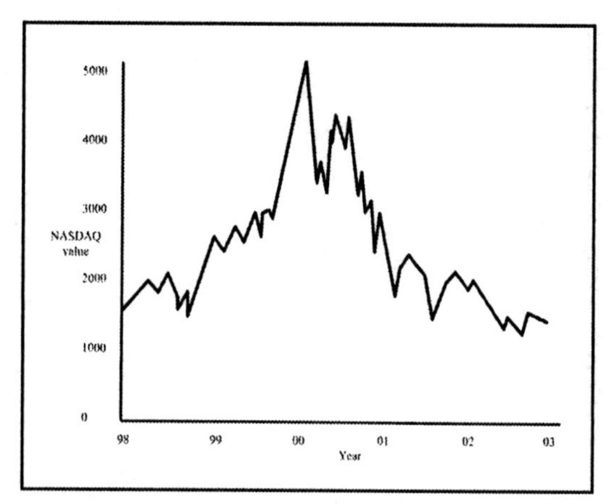

Figure 58. *How NASDAQ share values, biased towards technology shares rose and then collapsed.*

In first three years of the new Millennium, the bubble burst. The majority of dot.coms collapsed and as a result, stock market values plummeted. It is useful for anyone looking to venture into the area of the mobile Internet to understand why so many dot.coms failed to perform. The main reasons were that:

- Inexperienced leaders sought rapid growth as opposed to long-term success.
- Business plans were built on bright ideas and network effects rather than profit.

- Lack of an established business model. Assumptions were made that an Internet presence in itself would lead to profit
- Too many dot.coms - all aiming to monopolise the same market sectors.
- Too much money was spent, particularly trying to build market share.
- When dot.com failures grew, investors simply ran out of capital to invest.

There are several lessons to be learned that are equally applicable to the mobile sector. If you are going to grow a company you will need an experienced management team, often difficult to achieve in a new technology area. You will have to establish a viable business plan showing a path to profit based on a proven business model. Furthermore, you must recognise that too-rapid growth is unsustainable and that buying market share at the expense of profitability cannot be protracted.

In addition to the dot.coms, a number of communication companies had expanded rapidly and, as a result, were burdened with impossible debts. Many were forced to sell assets or file for bankruptcy. Worldcom, overstated its profits by billions of dollars. Following a crash in its share value it became the largest ever US corporate bankruptcy. Other communications companies, such as Marconi in the UK, got into similar financial difficulties as the forecast demand for higher-speed infrastructures failed to occur.

Yet some dot.coms survived and have become hugely successful. Examples include Amazon, Yahoo, eBay, Google and Expedia, although Google and Yahoo in particular have had to change their business models several times over the course of their development. Amazon and eBay, and to a lesser extent Expedia, have succeeded primarily by dominating their chosen markets. Many conventional companies have also successfully grown the Internet's share of their business, but the common factor behind all of the successes has been a focus less on clever technology and more on successfully leveraging money out of their users.

At the height of the boom in 2000, AOL had acquired Time Warner, the world's largest media company - yet within a couple of years, both the CEOs who made the deal had left. By 2003, AOL Time Warner removed AOL from its title. The dot.com boom was finally over and analysts once more acknowledged the importance of traditional business plans.

There are some similarities with what happened when British, German and Italian mobile operators put themselves deeply in debt by spending billions on 3G-licences. The UK licencees, Hutchison, Vodafone, O_2, T-mobile and Orange, between them invested over £20 billion in acquiring their licences. Given the annual profits from 2G licenses, it is likely that it will take some five years even to cover the license cost. But for these companies long-term survival is more important than maximising profit.

Mobile browsing

After the appeal of the Internet became clear, it was inevitable that when mobile phones had suitable colour screens and the necessary technical capabilities, users would find Internet access on the move an enormous draw. Some functionality was particularly suitable: access to travel timetables and near real-time arrival information; the ability to see theatre, cinema and concert programmes; choosing and booking seats; the capacity to follow eBay count-down times. All of these would prove almost irresistible attractions for people on the move. However, everything was and still is dependent on website owners making them suitable for viewing

Internet developers need only design web pages to support the most widely used browsers, Internet Explorer and Firefox. However on mobile phones there are a wide array of different browsers, mobile phones with different screen sizes, and beyond that there are three differing standards for how to develop pages for mobile phones: WAP and xHTML in Europe and the US, and i-mode in Japan. In Japan progress has been rapid and the majority of Japanese websites have been converted for mobile phone use, but the same is not true in the rest of the world.

i-mode is good

In 1997, a review of Japan's PDC digital network, suggested that there was insufficient capacity in the network to support expected growth in usage over the next four years. At the same time, PC penetration was very low in Japan and dial-up Internet access was very expensive. NTT DoCoMo decided that the solution to its problems was to get its customers to use more data instead of voice calls on its existing network.

With the WAP protocol discussions dragging on, NTT DoCoMo developed its own solution, i-mode, on available technology that would do the job and could quickly enter service even if it provided a less elegant solution than that planned for WAP. An entirely new infrastructure was built and i-mode, launched early in 1999, the world's first mobile packet-based data service for 2G phones.

It proved incredibly popular with Japanese consumers and enabled NTT DoCoMo to continue to add functionality, to the extent that the company had a virtual monopoly on the phone business in Japan in the early years of the new Millennium. NTT DoCoMo avoided both network saturation and the loss of customers to other network operators, while at the same time developing a protocol that proved more successful than WAP and could subsequently be licensed internationally to other operators.

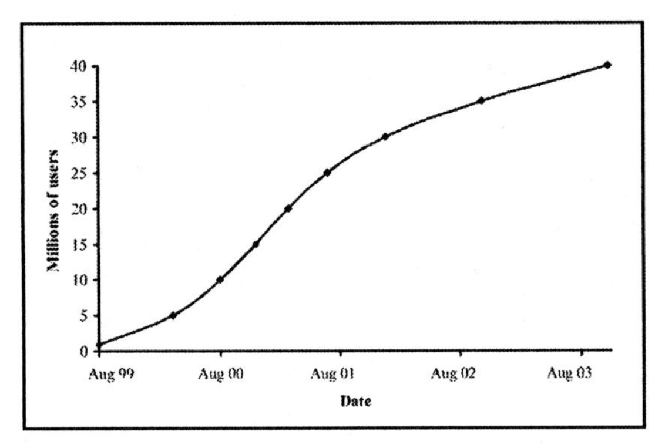

Figure 59. *The growth in numbers of i-mode phones over their first four years.*

What is i-mode? It is a solution that allows users a range of facilities including the ability to send e-mail, surf the Internet, check the latest news, play games, shop and book airline tickets. The i-mode display allows users to obtain information from i-mode menu sites and compatible Internet sites, and exchange i-mode mail in a very similar way to a conventional web browser.

Within fifteen months, 70% of i-mode users were males who also accessed the Internet at work. Half of them were still in their twenties with another third only a decade older. Work to attract female users succeeded and by mid-2001, 30% of Japanese mobile phone owners were using i-mode. At that time some fifty million

Japanese were estimated to have mobile Internet access. Four years later this number had risen to approaching eighty million, representing 60% of the population. Furthermore, around three-quarters of Japanese mobile phone users access the Internet on the phone every day. Among the Japanese who want travel information, a recent survey shows that just over half look to the Internet compared to one third using a travel agent. So why does Japan have so many users?

i-mode was designed from scratch to be user friendly and displays websites well, defining what appears on the screen using a mark-up language that is a backwards compatible subset of the language used by the world wide web: HTML. It provides services to users more quickly than WAP phones can. The virtual monopoly of i-mode in Japan plus the ease of developing sites, meant that it was very cost effective for owners to develop their i-mode sites. For end-users, the service has a simple sign up procedure and the cost is modest. Users pay only for the amount of data transmitted rather than the time they are connected to the web, although there are extra charges for some services that are billed through a built-in micro-payment system.

A factor that simplifies Internet access in Japan is that mobile operators there interact very closely with handset manufacturers. Mobile phones are developed to meet particular requirements; all have colour browsers and can run Java applications. This means that unlike Europe, they don't have to deal with a wide variety of keyboard layouts, screen sizes and pixel numbers. Most Japanese phones provide a standard way for users to access the Internet and NTT DoCoMo's huge market share ensures uniformity in the handset market and a fair degree of clout with handset manufacturers. Contrast this with the situation in Europe, where there are hundreds of types of handset making the question of Internet use harder both for network designers and mobile users.

i-mode tends to be equally divided between e-mail and Internet browsing, the latter including downloading applications, audio and video. The most popular sites in Japan are those providing news, including sports news and weather, games, online banking and share trading, travel and restaurant bookings, chat lines and personal home-page creation. Adult content pages also prove extremely popular, and the ever present sites to download ringtones and wallpapers are also there, but not as dominant as they are in Europe.

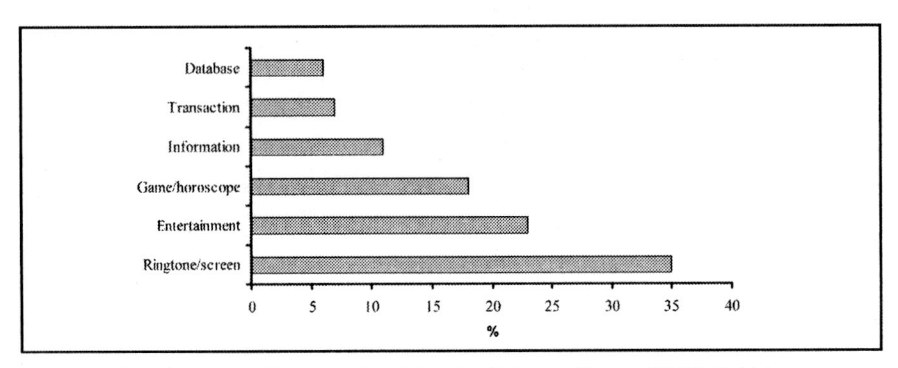

Figure 60. *The eight most popular categories of download in Japan. (Source: NTT DoCoMo)*

The i-mode business model is based on a modest monthly subscription plus a similar fee for each premium-content service selected by the user with a charge for each packet of data transferred; relatively inexpensive for small downloads like ringtones but still a disincentive for large ones. NTT DoCoMo charge users directly for the packets and provide a central payment system for content providers, taking a fee amounting to just 9% of each contributor's revenue; the remainder is shared between content owners and aggregators, advertising companies and new application developers. This is significantly more generous than revenue shares offered in other parts of the world, and most content[10] providers cite this as the key factor in the success of i-mode.

All this has occurred in a market where handset standardisation and the extensive understanding of the programming language used by the service have been crucial. The culture of the Japanese has also been a factor. The majority of the population is on the move and spends a lot of time commuting in a cramped environment that lacks the space to use laptops. In addition, the Japanese have not embraced the use of home PCs to the same extent as Americans and Europeans. Mobile Internet access was clearly what the Japanese needed and they took to it when it came in numbers not imagined or replicated in the West.

[10] Given that many mobile applications miss their revenue targets by factors of ten or more, it is not clear that a three-times increase in revenue share would really improve their business that significantly.

Indeed a lot of public services are offered over i-mode rather than the web: there is an interesting i-mode disaster message board service in both Japanese and English that is activated in the event of a catastrophe such as a major earthquake. It allows i-mode subscribers within the disaster area to post messages to family and friends regarding their safety and location. The service aims to prevent network-performance degradation due to an overload of voice and data following a major disaster.

Since 2000, NTT DoCoMo has been attempting to licence i-mode to network operators in other countries, and the fact that i-mode is not an open standard did not help its cause. It is only now that i-mode is being rolled out across Europe. Unfortunately, with WAP now widespread and with xHTML emerging as the second generation standard it is unlikely that i-mode will gain significant penetration. Most content providers want to support another standard like a hole in the head, although the increased revenue share offered by DoCoMo, if copied across Europe, may yet swing the balance in its favour.

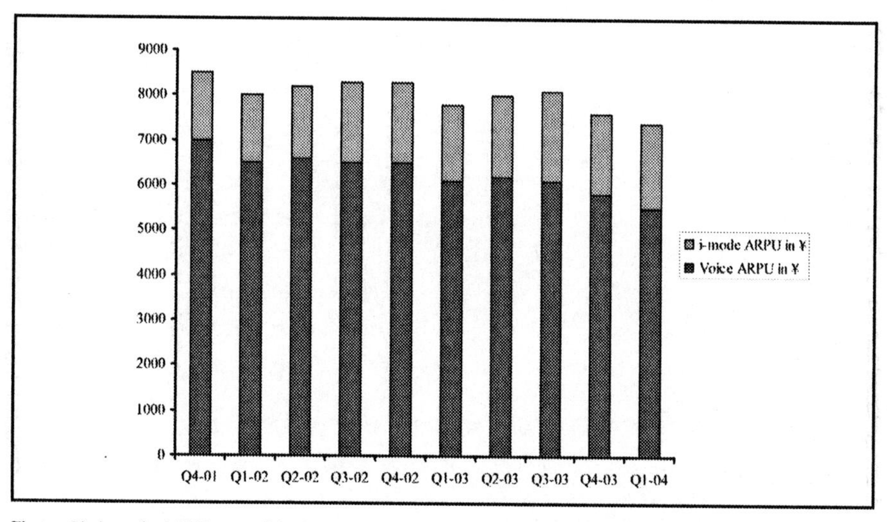

Figure 61. *i-mode ARPU growth in Japan was rapid in the early part of the twenty-first century. (Source NTT DoCoMo)*

Case study: CYBIRD content aggregator

A mobile phone isn't very exciting for users without an abundance of diverse, engaging, informative, and fun content. That is the driving philosophy behind the leading Japanese mobile content aggregator, CYBIRD's ongoing efforts to collect an extensive, library of wireless content appealing to myriad users and running on a wide variety of devices and networks in Japan.

The company was quick to recognize the demand by wireless users for very specialised information. One application, Namidensetsu (surfing information), for example, became an early and long-lasting hit. Namidensetsu provides wave and weather information about two hundred Japanese and forty international surf spots. In addition, it offers instructional information, surf-shop and tour information, advice for bodyboarders, and a service for downloading popular surf brand logos. A similar service is available for fishing buffs.

CYBIRD offers an information and communication site for girls and young women in their teens and early twenties. It's called Popteen and it offers fashion information, astrology and news about celebrities. CYBIRD also has built wireless online communities for cat lovers and women office workers.

85% of CYBIRD's revenue comes from content service subscription fees. These fees typically are billed through the carriers. NTT DoCoMo, for example, bills subscribers and keeps 9% of the fee for itself, passing the remainder to CYBIRD. CYBIRD also provides marketing solutions for businesses, which generates 12% of its revenue. International business consulting accounts for the remainder of its revenue.

Although CYBIRD develops some content itself, it acts mainly as a content aggregator, acquiring most of its content from independent content creators. It pays these content originators about 30% of the subscription fees. After hosting and transaction costs, CYBIRD retains about 50% of the subscription fees as profit, according to published reports. The company offers both consumer and business content.

The key to CYBIRD's success is its content acquisition. The company follows a five-step process for acquiring and developing content:

1. Identify appropriate content and negotiate with the content creator for the right to use that content in the mobile market.

2. Plan and design the content to fit the mobile format.

3. Solicit mobile operators for the content to secure appropriate placement of the content on the operator's mobile portal.

4. Develop the content for the operator's network and specific devices.

5. Initiate the service.

To ensure quality, CYBIRD supports a full department to handle application testing and a 24/7 mobile operations centre to provide customer support.

The problem of keeping content up to date is addressed from the very earliest stages in the process. 'When we plan new content services, we make a budget and schedule for updating and revisions,' explains Shinji Terada, vice president of the Strategic Technology Planning Department. Once content is in production, it is checked daily by the operations centre. The company also considers user requests and other market analysis when updating content.

CYBIRD has also created two utilities to facilitate access to online content – a sound preview (audition player) for its Cool Sound ringtone-content service, and a thumbnail viewer for its Cool Screen wallpaper-content service. Without such utilities, users who want to select either a ringtone or a wallpaper must repeatedly download each offering to find their chosen tone or wallpaper. This wastes users' time while running up online minutes. With the CYBIRD wallpaper utility, users can view multiple sample thumbnail images at a glance. The utilities proved so popular that CYBIRD now licenses them to other companies.

Source: Forum Nokia

WAP ain't so bad

In retrospect, NTT made an excellent decision by parting ways from the WAP standard. WAP is one of the few examples of a bad technology decision having a significant impact on the business. By the middle of 2001, Internet-enabled WAP phones only accounted for some 10% of all mobile phones and there were virtually no services available except on the main operator portals.

The operator portals continue to this day to dominate the WAP world, but this is more an indication of the shortcomings of the technology than the power of the operators. There are three basic problems with WAP:

1. It requires a very different mark-up language to the Internet. WML is not backward compatible with HTML and it can be complex to convert standard websites to WAP.

2. There is no standardisation on the browsers used for WAP and the form factors of mobile devices range from small black and white screens to large full-colour ones.

3. There is no standard for microcharging on WAP and content providers need to use either premium SMS or work directly with each of the different mobile operators.

Although the first of these problems has been removed by the emergence of xHTML as a successor for WML across most modern handsets, the issue of charging is outstanding. The launch of new services outside the operator portals is frequently confounded by a walled-garden approach taken by many of the operators to prevent unregulated content, and especially VoIP services (see Chapter 10) being launched without their approval.

Despite all this, browsing figures in the UK have continued to grow, and 2005 has shown some 25% increase over previous year's figures as the number of 3G phones starts gradually to climb. Monthly viewing figures now total around two billion page impressions per month. Part of the reason for this increase is the improvement in the ease with which mobile users can browse. This is partly due to the emergence of single-button browsing on the handset, but also to the availability of colour screens and more reliable data networks.

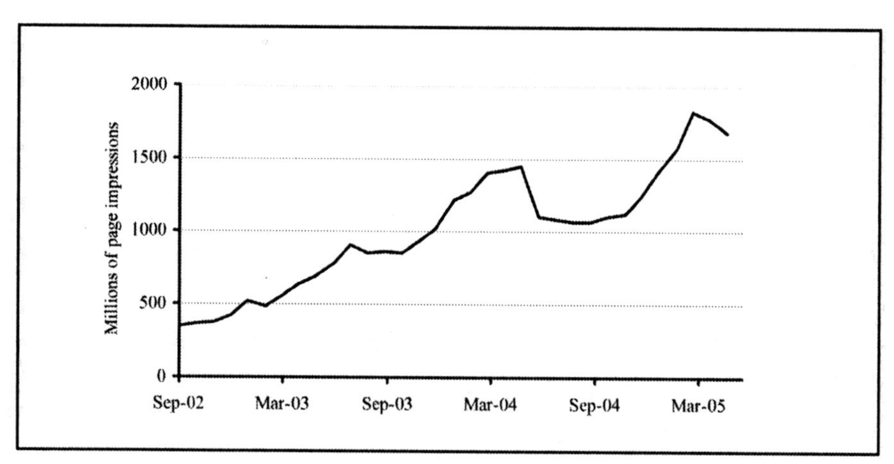

Figure 62. *The increase in WAP page impressions in the UK 2002 – 2005. (Source: Mobile Data Association MDA)*

If you are offering services over WAP, your primary route to market should be through the operator portals, although be aware that this is likely to change over the next two years. There are two ways you can monetise your service:

1. Pay-per-use – Consumers pay each time a service is used or content downloaded. It is clear exactly how much they have to pay, but it may bias them against regular use of the service.

2. Pay-per-subscription – Consumers are charged a fee for a pre-defined period to access sites. This helps to persuade users to make regular site visits as there is no cost disincentive.

Consumers are of course charged by the operator for the data-access fees, but operators are unwilling to share this revenue with content providers. Consumers will typically pay about £1 per Mb download, and you should factor this into your pricing considerations when setting an attractive price for your customers.

Price is, of course, only one of the factors in the data market, following the lead set for large free bundles of mobile voice minutes and text messages. Users are wary about incurring charges that are not included in their fixed monthly fee, but all of the evidence suggests that the real issue is a broad lack of compelling content for mobile devices.

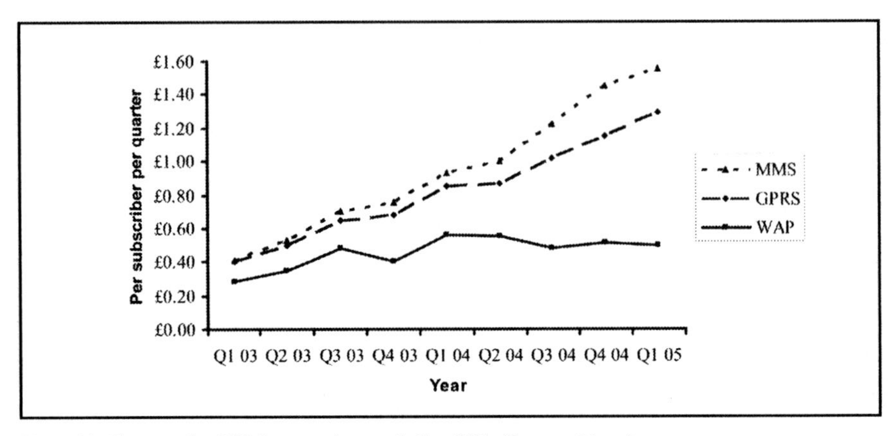

Figure 63. *The growth of UK data services excluding SMS. (Source: Ofcom)*

Case study: Vodafone 3G customers get Find and Seek

Mobile Commerce acts as a mobile content aggregator and distributor, enabling companies to deliver content to mobile users. This is possible through the company's best-of-breed mobile-content-delivery platform. Mobile Commerce is also the UK's leading provider of location-based services and has updated and extended the Vodafone Live! `Find and Seek` service as well as launching a 3G version, which features Multimap's content-rich maps.

The updates allow Vodafone customers to organise activities from cosying up to Tom Cruise in their preferred cinema, to locating a sports bar to watch the world cup qualifiers – all direct from their mobile phone. In total, Mobile Commerce has enhanced the 'Find and Seek' offering with an additional thirty-one services ranging from clubbing information to hotel reservations.

The service, available to 3G customers, is a very visual experience. Users who select 'Find and Seek' from the Live! menu are presented with a map based on their current location. Users are then presented with a list of five service types in the vicinity. The selection of services presented to users is dependent upon the time of day so, in the evening services include bars, restaurants and cinemas, whereas services promoted during the day include services like coffee shops. From this screen, users select the service they require and are then provided with more details. So, for example, if they wish to see a film, they select 'cinema.' The service then locates the nearest cinema and advises the user of show times and prices for the latest release. To help users who are having trouble choosing between films, the service also provides access to reviews and even trailers, bringing 3G customers closer than ever before to their favourite stars of the big screen.

"'Find and Seek' has proved to be an exceptionally popular Vodafone Live! service. Our customers really value having access to a wide variety of information relating to social activities like comedy clubs and clubbing guides. 'Find and Seek' allows them to view all the information that they require at their finger tips. Being able to view venues' exact whereabouts on a detailed map is signalled by users as a particularly useful tool, as location is frequently a key decision-making factor when it comes to entertainment decisions," said Al Russell, Head of Content Services, Vodafone UK.

"Through our partnership with Mobile Commerce, we recently extended the number of content providers supplying data to 'Find and Seek' so now our customers have access to more information than ever before. An additional plus for us is that Mobile Commerce has co-ordinated and managed these new relationships on our behalf."

Vodafone customers who don't have 3G handsets can still access the new content and services which deliver a dynamic and engaging service to users. For example, when logging onto the 'Eat' section, customers can choose from special offers, city eats and local restaurants. They can drill down further to select a particular cuisine, pricing option, restaurant-opening times and other facilities. The user can then select a restaurant and access walking directions of how to get there.

Content on 'Find and Seek' is now supplied by ten different content providers including Itchy, Toptable, Press Association, Ticketmaster, FilmNight, Active Hotels and Ents24, providing users with a more diverse range of services. The 'Find and Seek' service is hosted and delivered to users via Mobile Commerce's content application platform.,

Source: 3G Forum

The lack of take-up of WAP has led to some opportunities in the market, which have quickly been seized upon. It is well known that the killer application for the Internet is e-mail, and in Japan this is the most common use of i-mode. The lack of any convincing way to pick-up e-mail on regular mobile phones has left the way open for Research In Motion's BlackBerry product to take a dominant position in this market.

The BlackBerry is a multi-purpose hand-held device that wirelessly accesses existing e-mail, including attachments, provides mobile phone facilities including voice and SMS, an Internet browser and a calendar, address book, memo pad and task list. Essentially it is Microsoft Outlook in your pocket, designed from the beginning with e-mail in mind, with a QWERTY keyboard rather than the usual number pad. This makes it difficult to use as a phone and users frequently have separate devices for e-mail and voice.

Figure 64. *A BlackBerry with QWERTY keyboard and an e-mail.*

Many businesses have rolled out BlackBerrys as the mobile e-mail solution for their overworked executives. However, it is unclear whether the separate device will stand the test of time as the mobile Internet becomes more mainstream. Nokia have recently launched their first device supporting BlackBerry software as a client application, and it is likely that as the mobile Internet matures, Research In Motion will become more of a software company than a niche hardware manufacturer.

Location-based services

A lot of the promise of the mobile Internet has been based on the idea that the mobile phone should 'know' where it is and as such mobile Internet services can be more closely customised to the surrounding area.

By aligning wireless content information with location technology, relevant and significant location-based services can be made available. These services can be used for emergency purposes as well as business applications such as traffic updates, fleet management and asset and people tracking. L-Commerce is location-based e-commerce that responds to a user's actual location and has the potential for retailers to text offers to users when they are seen to be in the vicinity of their outlets.

There are various techniques for establishing location, starting with the basic cell of origin method[11], which identifies from which cell the call is being made. It requires only minor infrastructure changes by the network operator. The time difference, time and angle of arrival methods all require data from the network's radio masts and provide rather better accuracy. At the other extreme, by integrating GPS with a mobile phone, the use of GPS data provides a high degree of positional accuracy.

Unfortunately, many networks have had significant problems rolling out reliable location based services. Partly as a result of this, the FCC in America has mandated that future generations of phones should include a GPS receiver and that this location information is transmitted in calls to the emergency services. Although this mandate applies only to the USA, it is clear that this will drive the low cost production of such technologies and it is predicted that a very high percentage of mobiles will come with GPS within the next five years. This should kick-start the development of mobile applications that use location to tailor the information that is presented to the user.

Case study: 3 Hong Kong

3 Hong Kong has been quick to introduce a broad range of location based services using integrated GPS in the handset, including:

1. Display of a local map with the user's position on it.
2. Friend tracker to find other people by entering their phone number.

[11] Remember that mobile phone technology is still a cellular-based system.

3. Search engine for local services based on the user's current position.

4. Route-finder.

These sophisticated services are proving popular with users and it is likely that they will be developed significantly over the coming years.

The future of mobile browsing

There are three factors that will drive the take-up of mobile browsing in Europe:

1. More reliable, faster, networks.
2. More sophisticated devices that are capable of displaying content more closely aligned with the World Wide Web.
3. Significant entry into the mobile content market by major publishers with compelling content.

However, it is really only realistic for major publishers to get involved on a large scale once the first two technical obstacles have been overcome. The first factor to consider is the roll-out of next generation networks. These are covered in more detail in Chapter 10, but needless to say the faster networks will initially be unreliable and it will probably take three to five years before Europe and the US have widespread mobile broadband.

In order to overcome the limitations of existing networks, there are a number of companies developing thick-client applications that will provide a better user experience than traditional browsing. Leading the pack are Macromedia with their FlashLite product set which allows the rapid development of mobile phone applications, comparable with their web products. There are various other start-ups offering products in this area, and it is likely that these businesses will flourish. It is also worthwhile to consider client application development as a way to improve the user experience in the short term. However, do not expect the development to be cheap as there are few standards in this area and you will need to develop different applications for different phones.

Even with a client application you are still limited by the form of the mobile phone. As phones have developed they have become smaller and can easily fit into any pocket or handbag, with the results that the screens are usually small as well. This limits the possibilities for mobile browsing, and although there is now a clear trend for larger screens it is difficult to see mass take-up of larger mobile phones just so people can have screens to browse the web. One development that could change this is the promise of electronic paper. For many years there have been demonstrations of foldable screens. A pull-out sheet of electronic paper certainly has the ability to revolutionise the way that people consume content on their mobile phones. The first e-paper publications are being launched at the end of 2006, so it is probably realistic to expect to see these integrated into phones in the next five years, but mass take-up is clearly ten to fifteen years away.

Conclusions

1. Size and use of the Internet continues to grow but its mobile use, outside Japan, is only just starting to show similar signs of growth.
2. Expanding the range of WAP pages is critical.
3. Growing availability of broadband on mobile phones looks like a sine-qua-non for serious advances in the mobile field.
4. Cost of data downloads is still an issue for many consumers but operators are starting to provide imaginative pricing models.

Chapter 9
Mobile media

> *"I'm too shy to express my sexual needs except over the phone to people I don't know."* – **Garry Shandling**

Beyond the phenomenal growth of ringtones, many businesses have significant ambitions for the sales of media on mobile phones. With the exceptions of adult content and games, the majority of these businesses have not managed to make any profit from their activities in this area. Anyone considering entry into this marketplace should consider very carefully the options available to them.

On launching Virgin Mobile in 1999, Richard Branson said "Our vision for the future is for people to book air and rail tickets, find out information, buy music, financial services, books and whatever they choose while online and on the move. And we want to link customers to other Virgin services through multi-media via the mobile network."

Six years later, his business has become the largest MVNO in the UK with large profits derived solely from voice and messaging, with a negligible media component. While it is certainly true that people will be doing all of these things on their mobile phones at some point in the future, it is exceedingly unclear when these products will become mass market, and what business models will succeed in these areas.

Gaming

Playing games on mobile phones has been popular right from day one. In 2000, Nokia's Snake was by far the most popular computer game in the world, with tens of millions of players in five different continents. Snake was a very basic game and only proved popular because it gave people something to pass the time in a free moment: sitting on the train, waiting for a friend, or just sitting around at home. Of course, Snake was also highly playable, fiendishly addictive, and simple enough that anyone could learn how to play it.

Subsequent generations of mobile phones have come with more sophisticated games pre-installed, but it was not until mobile Java took off that the games industry started to get involved on a large scale. Java for mobiles had first been introduced in 1999 but it was not until MIDP (Mobile Information Device Profile) phones became widely available in 2002 that high-end games became possible on mobile devices.

Many of the more successful mobile games were tied to big console game franchises, such as Splinter Cell or Prince of Persia, or well-known arcade games from the 1980s, like Space Invaders or Asteroids. The number of people playing games on their mobiles grew to the extent that during 2005, Ubisoft, the French games developer, sold as many games for mobile phones as it did for all of the other platforms put together.

The success of mobile games is partly due to the convenience of being able to play a game in a bored moment, but the simple, short playing times of these games in comparison with the complexity of most console games. While a console game is usually designed with a playing time of forty-plus hours, a mobile game can usually be finished in less than an hour. This means that people who simply do not have the time or patience to struggle with console games can still play, and enjoy, the simpler games on their mobile phones. The costs of development are also significantly cheaper, with mobile games frequently costing a fraction of a percentage of the cost of a big-name console game.

Most games are sold on operator portals, alongside ringtones and wallpapers (See Chapter 7). If you own media content that could produce a compelling game, you could consider commissioning development for as little as £50-£100k. With profits from games running at between £1 and £2 per game sold, you would not need to sell that many in order to cover your development costs, particularly if your property has strong international appeal.

Case study: Addressing the mobile-gaming opportunity

Vodafone UK, with fourteen point one million customers, recognised the potential for mobile games early and decided to address the opportunity as part of its Vodafone live! proposition. Vodafone live! Games Arcade offers its customers a wide range of downloadable colour Java games, thus enabling Vodafone to leverage the enormous revenue potential of mobile gaming.

"The launch of Vodafone live! Games Arcade as a managed service solution demonstrates Vodafone UK's advanced thinking and openness to adopt new service delivery strategies that drive tangible benefits to the company, from both a business case perspective, as well as with respect to the technical solution." Atte Miettinen, Chief Marketing Officer, End2End.

Over the past two years, mobile gaming has emerged as one of the major drivers of mobile data, estimated to account for €1.3 billion in revenue globally in 2004, and growing to €6.7 billion in 2009. Downloadable games are expected to be the main driver for the popularity of mobile gaming, accounting for over 82% of mobile gaming revenues in 2007.

The popularity of mobile games is driven by the increasing capabilities of both mobile phones and the access networks, with the combination of Java and GPRS expected to drive downloadable games specifically. There are already an estimated thirty two million active users of downloadable games around the world, and this number is expected to grow to reach two hundred million active users in 2009.

Courtesy: End2End.

Attempts to raise the stakes in mobile gaming beyond casual use have not on the whole been successful: Nokia launched the N-Gage as a competitor to Nintendo's very popular Gameboy. It was not successful. Primarily its lack of success was because Nokia failed to secure a sufficiently wide range of gaming titles for the platform, and secondly because the cost of the product was almost twice that of the Gameboy. Nintendo have subsequently released their generation Gameboy DS and Sony have also entered the market with their PlayStation Portable (PSP).

Although Nokia continues to update the N-Gage it is unlikely that the company will be able to compete head-on in this market over the long-term. It is more likely that Nintendo and Sony will incorporate more phone-like functionality into future generations of their consoles. The Gameboy DS and PSP both include WiFi connections that can equally be used for multiplayer gaming between two devices, or surfing the Internet when a WiFi signal is present.

At the moment, it is not possible to run VoIP software like Skype on the PSP or DS, but most observers consider this to be only a matter of time. Once you can use your games console to make phone calls, the mobile operators are clearly going to want to be involved in this business. Nokia and other mobile manufacturers may be forced to reconsider their position in order to retain their pre-eminence amongst the mobile operators.

Mobile Music

Unlike gaming, music has been a common mobile medium for over twenty-five years. When the then chairman of Sony insisted on launching the Walkman despite surveys that suggested that there was no significant market, a legend was born. Sold in its millions, the early Walkman played cassette tapes, followed later by CD versions to cope with the then new medium. Generations of youth have become plugged into their own personal music world, and it was clear that mobile phone companies could benefit by incorporating musical functions into their handsets.

Nokia launched its first phone with an integrated FM radio around the turn of the Millennium. The product proved a hit particularly with women who, not wanting the extra weight of a Walkman, simply used the phone to provide them with music on the move. When the early MP3 players were launched, it was an obvious next step to build in MP3 functionality and from about 2004 onwards, many of the higher-end handsets could play music in this way.

However, the potential of MP3 phones has to date been eclipsed by the success of the iPod, Apple's standalone MP3 player, which has dominated the popular imagination with its stylish looks and ease of use. Apple have combined their consumer-electronics business with online-music retailing through their iTunes store and although they sell an average of only twenty songs per iPod, their sales are sufficient to make them the largest music retailer on the web.

The volume of sales compared to the number of devices suggests that most users are simply copying music from their CD collections straight onto their iPods, or in many cases copying music from other people's CD collections onto their iPods. It reveals a key part in any successful mobile music strategy: people primarily want to play music they already own, not buy new music. This is slightly at odds with the business models proposed by the record companies and many of the mobile operators, who aim to generate significant revenues from mobile music. But if people are not going to buy songs in significant numbers to play on their iPods, they are probably not going to do so on their phones.

The success of the iPod, and its ever diminishing size, is another factor that is likely to impact the success of music on mobiles. Motorola's launch of the iTunes-compatible ROKR phone in 2005 was entirely overshadowed by Apple's simultaneous launch of the iPod nano. Although the ROKR includes iTunes functionality, it is much less stylish than many other phones on the market and has not sold well. The fact that the ROKR is larger than both a small phone and the iPod nano also destroys any portability benefit of only having a single device. Indeed MP3 players are shrinking to the extent that they can already be worn around the neck and it is likely that the future MP3 player may be no larger than today's pair of headphones.

With this in mind, it suggests that there is little space for the widespread sale of mobile phone music players. Although this is broadly so, the mobile phone market place is so large that the segment which is interested in a convergent device is still significant. Sony Ericcson have recently launched a range of mobile phones with Walkman branding, integrating basic MP3-playing functions to greater sales success than previous Sony Ericcson models, but still significantly lower than Nokia's 6230 which includes MP3 functions but no special branding or device customisation.

What is most likely is that device manufacturers will continue to enhance their devices with music-related functionality, and an MP3 playing phone will be as ubiquitous as the camera phone. This does not mean that people will stop buying separate MP3 players, in the same way that they still buy digital cameras. But it does mean that while some people will only listen to MP3s on their phones, everyone will have the ability to do so.

Most people, who choose to listen to music on their phones, will probably copy music from their CD collections in much the same way as people use their iPods. Some people will buy music from iTunes or another online store and download it onto their phones. Other people may also buy music directly over their mobile phones from a mobile music store. The economics of fixed versus wireless networks are discussed more in Chapter 10, but suffice to say that it is unlikely that people will choose a mobile music store over an online store unless the price of wireless downloads is comparable to fixed-line costs.

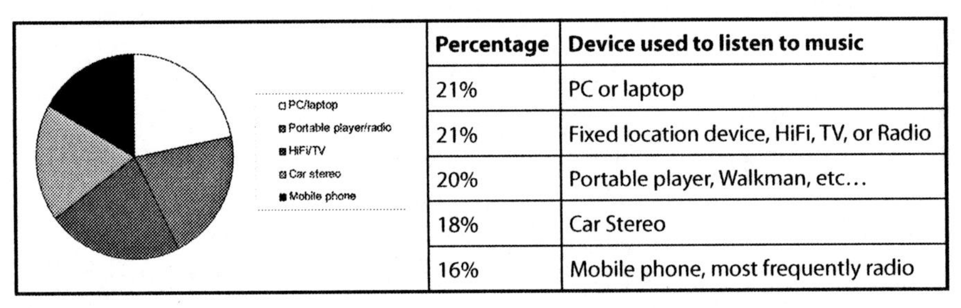

Percentage	Device used to listen to music
21%	PC or laptop
21%	Fixed location device, HiFi, TV, or Radio
20%	Portable player, Walkman, etc...
18%	Car Stereo
16%	Mobile phone, most frequently radio

Figure 65. *The popularity of the various devices on which listeners choose to hear music.*

Digital Rights Management (DRM) also complicates sales of music. When people buy music on CD, they can copy it onto their computer as an MP3. Because MP3 is a universal standard, music stored as an MP3 can be played on any different type of computer, phone or portable-music player. The problem of MP3 is that there is no way for the music companies to prevent people sharing their MP3 collections with all of their friends, thus illegally copying music. The solution to this problem is a collection of technologies that tie individual music tracks to individual computers or individuals to prevent people from illegally sharing music. Unfortunately there is no widespread standard for any of these technologies, and competing DRM standards mean that music purchased on iTunes will not play on a Sony MP3 player, and vice-versa.

This problem becomes worse if you listen to music on your phone, your PC and an MP3 player. Say you buy a track from iTunes. You can listen to it on your iPod and on your Mac or PC using the iTunes application. If you have a Motorola ROKR then you can listen to it on your phone, but not if you have a Nokia. Similar problems arise if you buy from Sony's online store, and these problems will undoubtedly slow down the growth of online music, particularly over mobiles where there are far more competing players than in the online music market.

DRM products are also hounding the conventional world of CDs, with Sony having recalled hundreds of thousands of CDs this year with faulty copy-protection software that opened up people's computers to virus attacks. Perhaps it is not surprising that people prefer to copy their existing CD collection to MP3 rather than buy new music online.

The one part of mobile music that is unlikely to be troubled by DRM is the transition beyond the FM radio tuner. In countries where Digital Radio has been launched, DAB compatible phones are already being released that support the massive increase in radio stations available, and with the roll-out of 3G networks, the growth in IP based radio is likely to grow. In 2005, the BBC broadcast a hundred and thirty four million hours of radio over the Internet. The BBC offers not only live radio but also on-demand repeats of the previous seven days' programming.

Of the opportunities presented by the mobiles to the music industry, by far the most attractive is the production and sales of ringtones. Consumption of music on mobile phones is likely to lag behind that of MP3 players and the business models for mobile music will most likely be based on the Internet using broadband music connections and then transferred to the mobile platform when flat-rate data access is widely available (Chapter 10). Any investments in streaming or music downloads are probably best made in the Internet environment for at least the next five years.

Mobile TV

Courtesy of the 3G networks, mobile TV is now becoming a feasible proposition. However, it is unclear whether it will represent an attractive proposition for mobile users; the screen is too small and probably the cost of watching for any length of time too high. The wider television market is moving towards ever larger TV screens in the home and it appears to many people that mobile TV is a step backwards.

Despite Apple's launch of the iTunes video shop, users are unlikely to be downloading full-length films and watching them on their mobiles. Downloading films is something relatively new even on the fixed-line Internet. The vast file size of a complete film and relatively slow download speeds had until recently, made this impractical. Today, with the speed improvements of broadband and new file-compression methods, film downloads are at last a practical proposition. Despite illegal file-swapping and free downloads, Hollywood film companies are making films available for public download and charging accordingly, following in the footsteps of the music companies. Whether this will prove popular with Internet users and be equally accepted by mobile phone users remains to be seen. There is the world of difference between watching a film on a laptop computer and watching one on a five centimetre wide mobile phone screen, and not a 'widescreen' at that.

However, there are a few niche areas where mobile TV does appear to provide a compelling proposition. Take particular notice of the propositions available when the content can be compressed into bite-sized two minute chunks. Previews of popular shows or trailers for new movies are already prevalent on the Internet and are easily downloadable over a 3G network. Sports-based content has been key in the uptake of SMS, and MMS video clips have hinted at the possibilities of real-time goal alerts. News updates can also easily be compressed into a short format that is readily viewable.

Longer formats are only likely to work when the content is simply not available in any other form. Live coverage of sports which are not sufficiently popular to justify broadcast on a TV channel and breaking news as it happens are probably the most obvious examples. Critical to the success of a mobile proposition is that people have an urgency to watch the footage on their mobile when it is not available elsewhere.

Case Study: Video Goal Alerts for EURO2004

T-Mobile was a major sponsor of UEFA's EURO2004 football tournament that saw national teams from across Europe competing for the prized trophy. One of the problems was that a number of the matches were broadcast on TV during working hours and viewers were often faced with the relatively unattractive prospect of taking time off work. As part of T-Mobile's sponsorship, T-Mobile had the right to broadcast video alerts of each goal to its customers as soon as the goal was scored.

The goals themselves were edited down into thirty second clips showing sufficient detail to identify the player who scored the goal and to see the action on a small-screen mobile phone. Users were charged a fixed cost per match or for the entire tournament and received videos for every goal in a game involving their national team.

The service was taken up across most of the countries with teams in the finals and proved that mobile video was not just possible on 3G networks but also on 2.5G GPRS.

In terms of the business models for distributing mobile TV, they are likely to follow broadly three different patterns, downloadable content, real-time broadcast and time-sensitive clips:

1. Downloadable content - Located from the operator portal or from off-the-page advertising. Video clips are sold in the same way as ringtones; downloaded to the phone and then played when people want to watch them. This model is likely to work well for promotional material such as trailers and pop videos which are not time sensitive, but does not work well for time-sensitive information.

2. Real-time broadcast - Most likely accessed either through the operator portal or through specialist 'TV' applications on the mobile phone. These give the user access to one of any number of channels that are streamed in real-time to the phone and are paid for either on a subscription basis or 'per-minute.' Real-time can also be accessed through a premium rate video-call, operating in the same way as traditional IVR, where the user watches the video rather than listening to a recorded message. Premium rate has the advantage of established business models and ease-of-use for the customer and is especially appropriate for off-the-page advertising.

3. Time-sensitive clips - Such as previews for tonight's episode of a favourite show. These are most likely to be promoted through off-the-page advertising in the listings magazines, with instructions to text a certain short-code to receive the video on your phone, or alternatively call a premium-rate video-line to receive the content.

Exactly which of these three models will be more successful and how much people will be willing to pay for them is as yet unclear. What is certain is that most broadcasters in countries with 3G will be looking to monetise their content onto the mobile platform. A combination of each of these different models is probably the best recipe for an initial trial.

Converging Mobile Devices

The distinctions between many different devices are already becoming blurred: at the end of 2005, there are some three billion mobile phones, two billion TVs and approaching half that number of PCs. You can already use your PC as a telephone, surf the web on your phone, and modern TVs have more in common with a computer monitor than a 1950s TV.

Nokia has established itself as the largest manufacturer of digital cameras. More camera phones are sold than conventional digital cameras. Apple is trailing far behind Nokia in terms of the number of MP3 players that have been shipped, but people are still buying iPods and digital cameras to use alongside their mobile phones.

It's clear that the way people consume media is changing, but the key question is:

"Which devices will people use, to consume which media?"

In all probability, people will use their mobile phones to consume some media, most likely that which is time-sensitive, urgent, or simply good to fill the time waiting for public transport. Whether films or TV programmes will prove watchable and attractive to mobile users is in doubt. Miniature colour TVs have been around for years but their sales have never taken off. On the other hand, the Walkman and the iPod have become icons for those who listen to music.

Whether the winners in the long term will be the mobile phone manufacturers or the consumer-electronics manufacturers is yet to be seen. Convergence is clearly on the horizon for most devices: perhaps more relevant is the question of whether people will be deterred by the relatively high price and inevitable complexity in the user interface of multi-function devices?

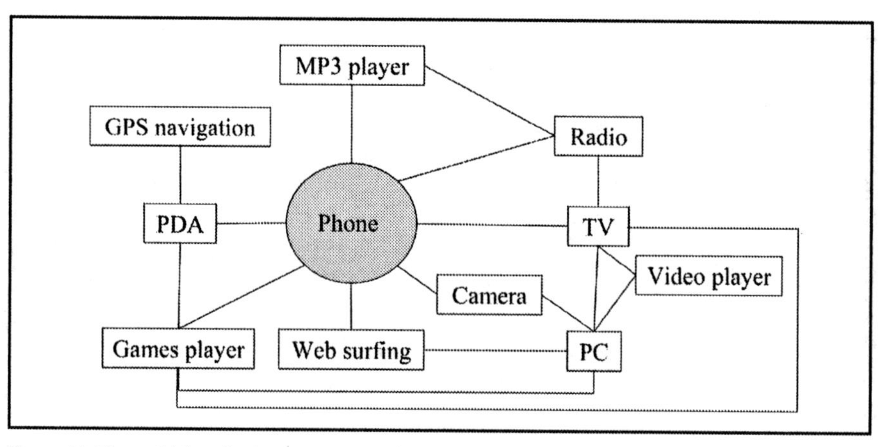

Figure 66. *The multi-function interconnections between the various mobile devices.*

Conclusions

1. The mobile media market should not be confused with the sales of ringtones and wallpapers that are primarily for handset personalisation, not media consumption.
2. Mobile gaming is currently the most successful of the mobile media markets, with sales close in numeric terms to conventional games sales.
3. Opportunities for music and TV on mobile phones are completely reliant on the roll-out of next generation networks.
4. The production of attractive, easy to use devices that can compete with the Apple iPod and other dedicated devices will probably be the turning point in the successful launch of mobile media services.

Chapter 10
3G and beyond

"For three days after death hair and fingernails continue to grow but phone calls taper off." - **Johnny Carson**

Many of the promises of the Internet have been fulfilled only with the emergence of broadband: with high speed access, flat pricing, and 'always-on' functionality, the Internet has moved from being a worrying cost that is slow and difficult to use to being firmly in the mainstream of society. Many people are hoping that 3G will do the same for mobile data, and although there are indications that the technology will be ready, there are several issues to be resolved before it truly becomes mainstream.

The limitations of 2G data services

The most successful second generation data technology is GPRS; initially developed as an add-on to the GSM standard for data communications. Previous methods were little more than running traditional modems over voice lines and resulted in very low bandwidth at a very high cost to the user.

GPRS took a different approach and a number of slots in the voice network are shared between all of the data users in a particular cell. This means that many more people can use the data network at any given time, but also that data is still competing for network bandwidth with voice. Voice tends to get priority and as such, during periods of high congestion, GPRS connections are unlikely to be reliable.

GPRS was the first widespread attempt to stimulate the mobile data market, and as such the manufacturers aggressively marketed it. Companies like Nortel hired people to travel the world evangelising about the capabilities of GPRS. On paper, it could achieve 128kb/s and was touted as the mobile ISDN. In reality the bandwidth was closer to 20kb/s – or the equivalent of a slow 56kb modem, and this failure to perform has led to many businesses becoming highly cynical about the possibilities of the Internet on mobile. Although other technologies emerged in other parts of the world, including CPDP and WiDEN in the US and PDC-P in Japan, all of them suffered from the same basic drawbacks:

1. The bandwidth is very limited compared to the regular Internet.

2. There is high latency on any connection.

3. There are long set-up times to connect to the network.

4. Configuring handsets appropriately seems to be highly problematic.

In Europe the limitations of the WAP infrastructure and unreliability of early WAP gateways compounded these shortcomings. Network management tools were also designed primarily for voice traffic and providing reliable network infrastructure for data connections was virtually impossible during the first few years after launch.

As the networks became more reliable, many of the intrinsic problems were overcome by specific packages of devices and custom-content services, such as BlackBerry and i-mode. The take-up has been very successful, but each of these packages requires specialised pre-configured devices, data-compression technology and custom content.

EDGE is a variation on GPRS that enhances the radio interface to dramatically improve the bandwidth available without the need for a wholesale upgrade to a 3G network. It has been rolled out significantly in the US and in a number of other markets where the move to 3G is likely to be slower.

The promise of 3G networks

Increased bandwidth is not the only reason to move to 3G. The quickest way to increase the bandwidth of a GPRS network is to roll-out EDGE. But this is not considered a full substitute for 3G.

The primary driver for the move to 3G networks is that they are technically better than their predecessors: they use bandwidth more efficiently, require lower levels of handset power consumption, and the associated licensing auction made more spectrum available for the mobile operators to carry their services. In the long term, the most important of these factors will be the availability of spectrum, as most of the proposed standards beyond the current 3G networks all use the same frequency range.

Of course, when the 3G licenses were auctioned and the business cases were being assembled, it was not sufficient to say that the systems were technically better. A number of 'killer applications' were proposed that suggested a dramatic improvement in the revenues of the network operators if they moved to 3G. The reality was that all of the operators had to move to 3G purely because their 2G services would no longer have been competitive once 3G became available. Basically, the driver to 3G was really a hygiene factor rather than a growth driver. Having said that, there are a number of specific features of 3G that are not present in 2G networks:

1. Much higher bandwidth, approaching that of broadband Internet. Certainly enough speed to download music but probably not enough to download high quality videos.

2. Video calling.

3. More reliable and faster Internet access and browsing.

There are now around four hundred 3G networks in service around the world, of which fewer than a third are in live operation. Initial reaction shows that they are able to deliver significantly more bandwidth than 2G networks, but that their early days are still plagued by reliability and handset availability problems.

A common theme amongst the improvements in mobile network technology is that it is usually accompanied by a backward step for the user in handset performance. When GSM handsets were becoming small and lightweight, GPRS was launched which required a new generation of larger, clunkier handsets. Early camera-phones were also large and it was only during 2005 that state of the art handsets of the size and reliability of 2001 devices became widely available.

Early reliability problems are normally accompanied by growth in usage. The old adage 'you always test in the real world' is certainly true when it comes to telecommunications and the reliability of 3G networks is likely to lag marginally behind their widespread availability.

Any business looking to launch services that require 3G capabilities should be wary of expecting significant revenues too soon after the launch of 3G, and an international strategy will be further confounded by different roll-out schedules across different countries.

Mirroring the Internet tipping point

3G has been operational in Japan for several years, and it is now clear what the 'killer applications' are in this market. Primarily, voice communication is still the main use for mobile phones. Second comes sending textual messages, increasingly as e-mails. Thirdly is picking up time-sensitive information like weather, news, and sports results. Downloading music has had a degree of success, although nothing comparable to the online phenomenon.

These are all basically extensions of the services that have been offered on 2.5G phones for a long period of time. Yet, it is interesting to note that the only revolutionary application, video calling, has to date been spectacularly unsuccessful in all regards, although there is great potential for premium rate video shortcodes.

Analysis: 3G in the UK

According to UK operator 3 in the first half of 2005, their ARPU in the UK was £33.83 with three point two million customers. That ARPU split between data and voice is fairly standard at 77% voice and 23% data. But they are paying around £180 per customer in acquisition costs and are still not breaking even.

This in many ways compares well with the incumbent operators: the average ARPU in the UK is about £24.50 overall or £27 for data users. However, subscriber acquisition costs are typically between £30 (Virgin) and £100 (Orange) and Vodafone has reported a 15% increase in ARPU for 3G customers which is similar to 3's premium.

This suggests that the early adopters of 3G are higher ARPU customers in general, and that 3 is benefiting from its aggressive marketing only because it is succeeding in attracting premium customers. Given that 3G users are still not spending a high proportion of revenues on data, voice is still the killer application for operators.

However, in launching 3G the operators are providing significantly more bang-per-buck in terms of data services. Combined with higher bandwidth and more PC-like devices, this does provide a huge opportunity for media owners to launch services on the mobile devices that can generate significant customer loyalty and revenues for them. As long as 3 is willing to pay a premium to attract customers, it is also likely the company will spend money with media owners to improve the perceived level of service they can offer in the marketplace.

The increased bandwidth and reliability of the data networks in 3G brings the opportunity of enabling a number of applications that have simply not been practical before:

1. High quality audio distribution (file sharing, MP3 downloads, VoIP).
2. Low quality video distribution (video calls, video clips, pop videos).
3. Extensive browsing with the use of graphics (web on your phone).

This makes 3G a basic pre-requisite for extensive take-up of these services across a wide user base.

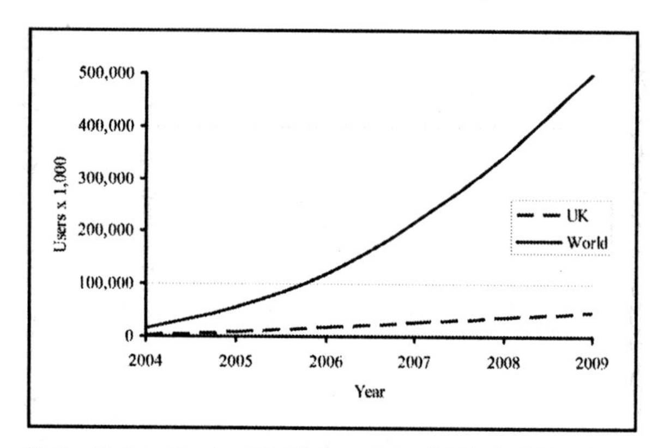

Figure 67. *An estimate of the likely growth of 3G in the UK and worldwide over a five year period. (Source: Ovum)*

The more optimistic forecasts estimate a good level of take-up by 2008. Note that these require massive increases in the number of subscribers during the next few years – certainly at a faster rate than the take-up of i-mode. It would not be surprising if the mass take-up required significantly longer than predicted.

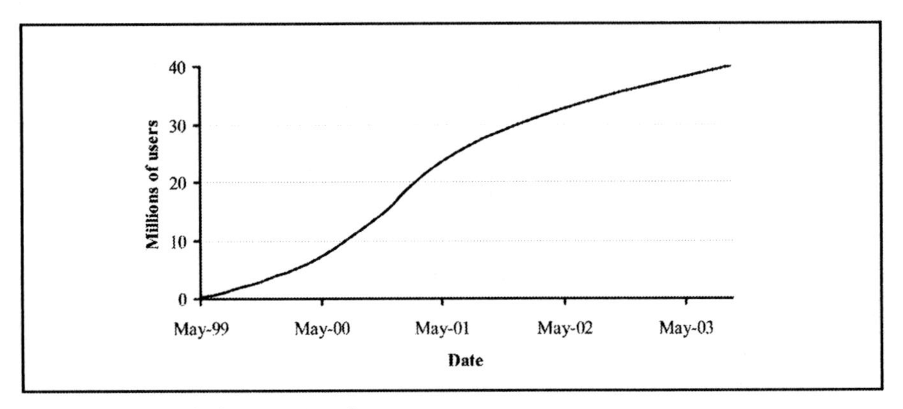

Figure 68. *The growth of Japanese i-mode use.*

The pricing problem

Many people outside the mobile industry – and some within it – consider that mobile data is prohibitively expensive and that the mobile operators, in adopting such a pricing policy, are guilty of significantly holding back the development of mobile data. It is certainly true that operators are deliberately pricing data above a usual market rate, but this is not to generally hold back mobile data, it is only to prevent the growth of (Voice over Internet Protocol) VoIP services.

Moving VoIP on to mobiles

VoIP is a method of phoning over the Internet and provides an interesting alternative to conventional landline phones and mobiles. A headset or microphone and speakers may connect to a PC or via a telephone, either with a built-in or separate VoIP adapter. These adapters connect into a customer's existing broadband modem. Companies like Skype, recently acquired by eBay, and Vonage are VoIP service providers, offering free calls over broadband connections to other users and low cost calls to conventional telephones, both in-country and abroad. Numerous IP service providers and telcos are rushing to offer VoIP.

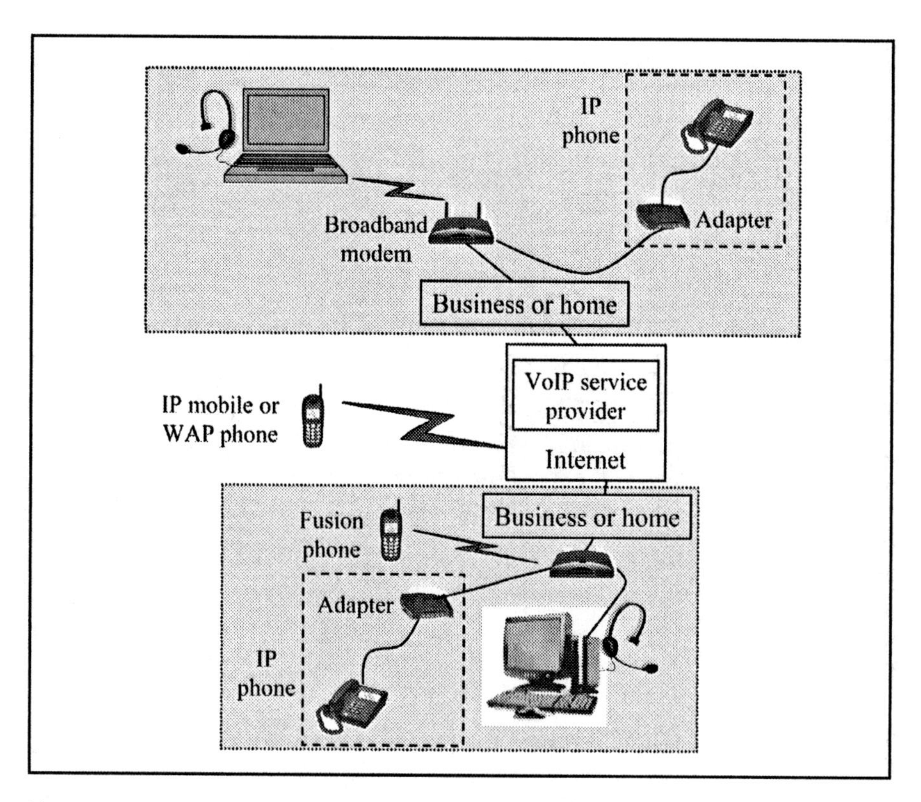

Figure 69. *How VoIP may be used by business or home users over fixed lines or with mobile phones*

On mobile networks, VoIP offers the potential for significantly cheaper calls that normal voice ones for the following reasons: firstly, they occupy less of the network operator's bandwidth; secondly they are not restricted by statutory charges like international accounting settlements and interconnect rates. Finally all that is needed is a connection to the Internet; normally free on fixed network and potentially inexpensive on mobile phones.

However, because mobile operators currently receive 80% of their revenues from voice charges, a widespread emergence of VoIP without a hefty chunk of revenue for the operator is simply not sustainable.

As we have seen in Chapter 4 the economics of running a network are primarily capital intensive with low variable costs of service; the mobile operator can survive equally well on £25 per month from its subscribers for data-only as it can on £25 per month for voice calls. The issue is a matter of migration. For many data services (such as mobile browsing) to be attractive to a user, the bandwidth must be priced such that VoIP can undercut traditional voice calls. Even if a small proportion of voice-calls move to the data network on a VoIP infrastructure, the capacity on the data networks would increase dramatically. This would give not only a reduction in revenues but also an operational issue for the networks to resolve.

All in all, it is clear that VoIP on mobiles will come: Nokia has already developed a special mobile IP phone and mobiles that run Windows CE can download Skype and use VoIP in the same way as any other PC user. It also seems that businesses and home users are starting to shift towards VoIP-based communications. The key driver is cost saving. The integration of mobiles, WiFi, and VoIP, has already started and is likely to become increasingly popular. Thus, almost every telco is attempting a converged fixed/ mobile service of some kind. As operators start to work more closely with VoIP providers, a migration from voice-centric pricing to data-centric pricing will appear that reflects more closely the fixed line ISP model – a fixed price per month for virtually unlimited data. Provided the average ARPU is still more than £25 per user, the mobile operators will be able to sustain the business of operating the network and we have a sustainable business model.

WiFi

WiFi networks are now widespread throughout homes, universities and offices in most of the developed world, as well as a number of extensive public-access networks in locations ranging from Starbucks coffee shops to airports. Some civic authorities (notably New Orleans) have even taken it upon themselves to provide free WiFi networks for all of their citizens.

Compared with 3G networks, WiFi is a high bandwidth, high reliability, radio network. The power consumption for WiFi is higher than 3G, but so is the throughput rate. However the range is short. This means that to provide continuous access to WiFi throughout a building will frequently require multiple WiFi access points. To cover an entire city, it has often been suggested that an access point should be installed on every lamppost.

Clearly this makes any widespread WiFi roll-out very expensive compared to other technologies, but given the success of the Starbucks model, it is certain that WiFi will have a niche in the networks of the future.

WiFi certainly has the bandwidth to run VoIP applications and the popular Skype program is frequently run by people over a WiFi network in their own homes. In many Starbucks, you will see people with laptops connected through to the Internet talking into headsets that use VoIP. The Starbucks' pricing is interesting – £5 for sixty minutes – which is not entirely dissimilar to the cost of a conventional phone call but of course the call itself is free, and the customer is just paying for the data access.

Beyond 3G

3G phones offer a significant increase in speed over previous generations, but the forecast demand for quality live TV on mobiles means that even higher speeds will be needed. 3.5G HSDPA can achieve transmission speeds of up to 14Mb/s that really offers broadband connection speeds to users, which is likely to become the standard that everyone wants.

Meanwhile China and Japan have decided to work together to develop 4G mobile telephony although the range of uses and technical details have yet to be decided. However, the networks will be faster – up to 100Mb/s probably using OFDM. This seems sensible since Japan was the first country to launch 3G and China is the world's largest mobile market.

While work on 4G mobiles is already in hand, it is unlikely that this latest generation of phones will start to come into widespread use before 2010. However, the new technology will certainly offer higher bandwidths and faster data-transfer speeds than previous generations of mobile phone.

While all of these standards have either been approved or are currently doing the rounds with the standards bodies, it is clear that there will be a similar level of fragmentation amongst each of these standards to those that occurred with 2G and 3G standards. This fragmentation will limit the interoperability between the different standards and increase the complexity required of the mobile device able to roam between different networks.

However, we must not underestimate the possible impact of the next generation of networks. The economics of running a wireless data network reduces with each subsequent generation and a time will come when it is cheaper to run a wireless network than a fixed line. Once wireless bandwidth is cheaper than fixed-line bandwidth, there is likely to be a growth in wireless-hosting infrastructure and a general move away from fixed lines.

The primary aim of the worldwide WiMax standard is to bridge the information and communications technologies gaps between the developing nations, where four-fifths of the world's population resides, and the already developed ones. It aspires to provide telephonic connection to the estimated million and a half villages presently unconnected to any network and grow the number of people with Internet access to at least half the population of the planet.

Many of the pioneers in the wireless field are now looking to completely obsolete fixed-line networks. There is a significant reliability curve that needs to be overcome before wireless networks have the same level of reliability as fixed. In the long term wireless is less prone to being disturbed by earthquakes, fire, terrorist attacks and other such events that frequently cause outages on fixed-line networks.

In addition, there is no need to dig up the streets to upgrade fibre, build new networks along the ground or under the sea, and many new wireless products can be upgraded much more quickly than their fixed line cousins. It is this speed of development, combined with the ever decreasing costs of running wireless networks that makes it likely that they will become the preferred choice for communications in the long term.

Case Study: Getting Broadband to the Scottish highlands

Demand for broadband in Scotland is high, as it is in the rest of the UK, but in the rugged mountainous regions there has inevitably been limited penetration of line-based services. A wireless solution can rapidly and economically deliver Internet services to users in remote locations despite the severe weather frequently found in the area.

The community broadband service offered by Dick Fleming Communications (DFC) meets the needs of remote groups who want broadband access. The system is based on a satellite receiver which provides the connection to the Internet and a wireless distribution system around the community into which users can connect by wireless. DFC supply, install and operate the entire infrastructure.

Ideally the system requires a minimum of twenty users to make it cost effective, although smaller groups can be supported. Customers require a computer with an Ethernet connection and obtain a connection giving surfing burst speeds of up to 1Mb/s, maximum download speed of 500kb/s and more than adequate access to the upload capacity. The service is not suitable for games due to the satellite latency of about one second, and the peer-to-peer capability is severely constrained.

Charges depend on the number of clients connecting. Above twenty the typical charges are a £150 one-off connection fee covering the client end equipment and then a charge of £25 per month. The system is also suitable for use by small businesses.

Courtesy: Dick Fleming Communications

Ubiquitous wireless

The 2.5G GSM/GPRS network is probably the last network that will be rolled out to almost universal coverage. It is a long-distance network that can provide high quality voice calls, and basic data services to remote areas at a low cost. From 3G and beyond, the effective range of the radio transmissions is such that future networks are only likely to be rolled out in more densely populated urban areas. It is most likely that future devices will be able to connect to many of the different available types of network, seamlessly (to the user) selecting the network with the best bandwidth or price, and providing a continuous connection to the Internet. Voice calls are more likely to be routed over VoIP connections when available and transferred to regular networks when the VoIP connection is not available.

If this sounds like a recipe for disaster in terms of technical complexity, it is worth noting that all 3G devices can connect back to 2G networks when the 3G is not available. BT in the UK has just launched its first convergent device. This phone uses a regular GSM connection for voice calls on the move, but then transfers to a Bluetooth connection when back in the home to route the call over the fixed line, providing wireless calling in the home for the same price as fixed-line calling.

In the future, it will not be a matter of considering whether you will call someone using Skype, your fixed line, or your mobile, but simply deciding to make a call and having your intelligent mobile device routing the call through the best available network. The same convergence is likely to occur with messaging: SMS text messages being used only on the most basic networks, with all other systems using either an instant messaging type of application or e-mail if the recipient is not available at the other end.

This has a significant impact on the possibilities available to offer services on mobile phones for non-mobile operators:

1. Who is going to manage roaming between the different networks?
2. How will bandwidth be used when these high speed networks become readily available?
3. As the highest speeds will never be ubiquitous, will this limit the services that will succeed in the minds of consumers?

The first question is really a matter of the evolution of the MVNO. With more and more different networks in place, there will be few, if any, businesses operating a full set of networks in any one country. The MVNO then has the opportunity to provide the branded device and billing relationship with the consumer who then roams freely from network to network.

If you are serious about your brand in the telephony arena, this must feature heavily in your ten-year roadmap, but how can you achieve it if you run your own network? If you're starting an MVNO, consider launching convergent devices early on, and setting up roaming agreements with providers of civic WiFi services and VoIP to get the best of all worlds.

Secondly, the amount of bandwidth is going to grow beyond anybody's wildest dreams. Forecasts show that HDTV broadcast over mobile networks should be possible within the next ten years. Does this mean that people will watch more TV on their phones or does it mean that TVs will become more phone-like? If TVs truly become phone-like, then it is certain that video calling from the TV (as shown in numerous science-fiction shows) will become a reality. Also, shows themselves will become even more interactive than they are now and the choice for consumers will explode beyond all demand. The question won't be: "What's on channel 5 tonight?" it will be: "Which episode of Friends do I want to watch tonight?"

Thirdly, the fact that these wireless networks will not be ubiquitous suggests that their applications will be less mobile. As is already clear, the success of mobile phones has been the result of their personal nature as much as their mobility. Integration of the address book into the phone has had sufficient impact to encourage people to use their mobiles at home when a cheaper alternative is readily available.

Truly broadband wireless networks will facilitate phenomenal bandwidth around urban areas, but little in the far corners of the earth. This is likely to exacerbate the digital divide. Critical services will never be available exclusively through high bandwidth networks: emergency services, public service broadcasting, and government communication will primarily stay on the ubiquitous networks, whilst only the highest end of consumers are able to get the most bandwidth.

Conclusions

1. The growth of 3G will critically affect future opportunities for services that require higher bandwidths to download music host video calling and provide fast, reliable Internet browsing.
2. Future networks will not be rolled out globally and the best bandwidth is likely to be available only in high-population conurbations.
3. VoIP will upset the current balance of the mobile industry and presents an opportunity for both new entrants and the mobile operators themselves.
4. Eventually, wireless will overtake fixed line for data usage and there is still significant growth for wireless-services providers.

Chapter 11
Backing the winners

"Prediction is very difficult, especially if it's about the future." – **Nils Bohr**

The world market for mobile phones is not a homogeneous one. There are significant differences between the US, Europe, Japan and the Far East, between the developing nations that are industrialising and those that are not. Japan is undoubtedly several years ahead of the rest of the world in terms of the sophistication of their mobile networks and the capabilities available to their customers.

However, as should have become clear from reading this book, on a global basis there are very few innovative business models that have succeeded in the mobile arena. The vast majority of successes are when businesses have taken business models that have been established in another area and applied them to mobile technology. Mobile phones have broadly been sold in the same ways as their fixed line predecessors, which have not in themselves developed significantly over the last one hundred years. Premium SMS has followed straight on from premium-rate phone lines. The success of ringtones and personalisation, is a simple extension of people being able to buy their own telephone. MVNO models have already been established in the field of billing for other utilities such as water and gas, and the Internet on mobile phones is rapidly converging on the same models as the fixed-line Internet.

When we start to look into the future it is less clear what will be successful, but it is likely that the business models will be based on practises that already exist in the fixed-line or non-telco industry.

In 2005, more than two hundred wireless start-ups were thought to be operating in the UK, generating annual revenues of some £2 billion; an average of £10 million per company. There is clearly money to be made out of start-up businesses in the mobile sector, and many of these are technology or service providers aligning themselves closely with the established players in the mobile industry in the hope of an eventual trade-sale.

For established businesses there are also significant opportunities. Convergence amongst businesses within the mobile value chain is creating openings for established brands to become significant new players in the mobile industry.

In the UK, BT is losing some of its traditional market, providing telephone and broadband line connections to businesses and homes and is looking to provide video and TV on-demand through a set-top box that also provides access to all the Freeview TV channels. Maybe it is also now regretting its de-merger of Cellnet, its mobile phone arm at the end of 2001 to become first O_2. The same may be true for AT&T as AT&T Wireless is now part of Cingular. eBay, the online auction business has entered the VoIP business, while Sky has purchased easyNet to expand into the ISP broadband market and agreed a partnership with Vodafone to provide mobile TV. At the same time, all media companies are recognising the drift away from conventional radio and TV towards PCs, MP3 players and mobile phones. And this is also having a negative impact on traditional advertising revenues.

In addition, communication services are converging. Paging facilities have been incorporated into mobile phones as a cost-effective way to receive messages. This is likely to limit the future of pager-only devices. PDAs are being merged with mobiles as are MP3 players and no doubt the same amalgamation will occur with the video iPod. In the living room, PCs running Windows Media Centre are expected to challenge TV/video/CD/DVD systems and Microsoft's latest dream is that working in conjunction with BT, it will fill a gap with Microsoft TV that will also allow viewers to send text messages via the TV. Even popular Internet content, such as blogs and avatars, are available on mobile phones and users can update blogs on the Internet via their mobile.

BT Fusion's new handset (currently a Motorola v560) works as a mobile phone when on the move but automatically connects to a BT broadband line at home. It gives the features of a mobile phone, and the usual mobile call rates, but at home the cost of its use is the normal fixed line rate. This is achieved by a

Bluetooth connection to a hub (a small box) at home that can also be used like a wireless router to other suitable devices in the home – computers, printers and the like. BT is planning to increase the capabilities of its Fusion system to provide VoIP calls. Its two minor shortcomings are that texts are always sent over the mobile network and the bill is not integrated with the BT landline bill.

Convergence is also occurring between companies. Mobile phone operators, Internet service providers, broadcasters and major retailers are all eyeing each others' market sectors. This is following the high street pattern where supermarkets sell petrol and petrol stations offer a range of food. Already, Carphone Warehouse is offering a broadband Internet service to its customers and has purchased the fixed-line operator OneTel from Centrica. Mobile phone operators are starting to provide radio and TV on their handsets.

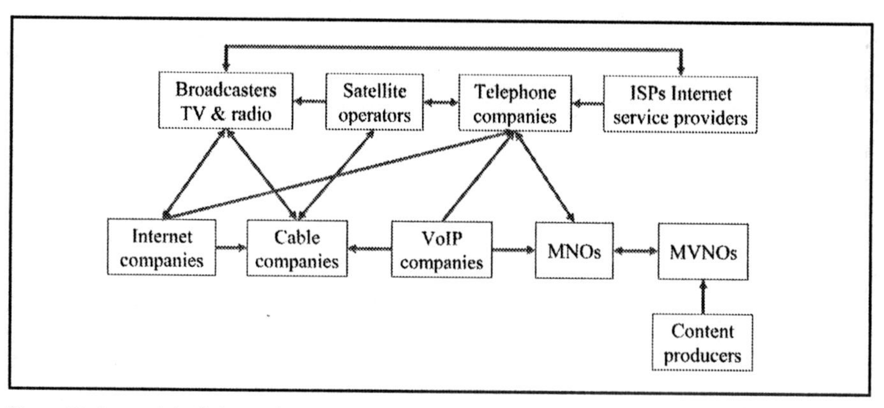

Figure 70. *Some of the linkages showing how different types of business are breaking into new areas.*

These shifts have the potential to create massive changes in the way that the telecommunications industry develops. At the end of the day, the opportunity for established media brands, start-up businesses, and retailers to benefit is immense. The different strategies you could pursue are briefly summarised in the table overleaf.

What strategy you pursue will really depend on the assets in your business and how they can be leveraged in mobile. It is certainly clear that there are some significant short-term revenue opportunities where most businesses should be able to benefit.

Strategy	Effort Required	Benefits
Start your own MVNO	Significant new product development	High revenues, increased customer loyalty
Launch SMS based services	Low product development costs	Strong revenues in the short term, less clear mid-long term
Use your brand assets for handset personalisation	Diversification of existing products	Minor revenues, increased customer loyalty
Launch products on the mobile Internet	Diversification of existing web properties	Little return in the short term. Potentially huge mid-term gains
Enable the explosion of mobile media	Significant product development required	High-risk investment with potentially high returns
Become a player in ubiquitous wireless	Very difficult for non-telcos to enter	Potentially massive returns in unknown timescale

The longer term investments are all more problematic and probably best left to the incumbent telcos – who must solve the problems for their business to continue – or the likes of venture capital funded start-ups which can tolerate a significantly higher level of risk than most established businesses.

Whatever strategy you undertake, take the time to learn what has gone before. Learn from their mistakes. Capitalise on their successes. The mobile path is littered with the casualties of new business models and products that are simply not viable. Focus on the established practises that have developed over the last one hundred years of telecommunications and you will find that success is not that difficult to achieve.

After all, if a Crazy Frog can make it to number one, how much better can you do?

Glossary

Many terms used in the mobile industry are subject to varied and vague definition, and this glossary should only be considered an authoritative guide on the meaning of terms when used in the text of this book, as the meaning in other contexts may differ.

1G The first generation of analogue mobile phone technologies including AMPS, TACS and NMT.

2G The second generation of digital mobile phone technologies including GSM, CDMA IS-95 and D-AMPS IS-136.

2.5G The enhancement of GSM which includes technologies such as GPRS.

3G The third generation of mobile phone technologies with increased bandwidth to enable multimedia applications, and advanced roaming features.

AMPS Advanced Mobile Phone System, the analogue mobile phone technology used in North and South America and in around thirty five other countries. Operates in the 800MHz band using FDMA technology.

Analogue The representation of information by a continuously variable physical quantity such as voltage.

ARPU Average Revenue Per User.

Bandwidth A term meaning both the width of a transmission channel in terms of Hertz and the maximum transmission speed in bits per second that it will support.

Bluetooth A specification designed to enable mobile phones, PDAs, computers and a wide range of other devices to share information and synchronise data. Bluetooth requires a transmitter/receiver chip in each device and will operate within a ten metre range.

Bit/s Bits per second.

Broadband An always on service that allows simultaneous use of voice and data services and provides a faster downstream connection than a dial-up service.

Capacity A measure of a mobile network's ability to support simultaneous calls.

CB Citizen's Band.

CD Compact Disk.

CDMA Code Division Multiple Access cellular systems utilise a single frequency band for all traffic, differentiating the individual transmissions by assigning them unique codes before transmission which are used to reassemble them on arrival. There are several variants of CDMA.

Cell The area covered by a mobile base station.

Circuit switching Creating a connection by opening a dedicated circuit between both parties to a call. The circuit remains open for the duration of the call.

Coverage The geographical reach of a mobile phone network or system.

D-AMPS Digital AMPS, a US wireless standard also known as IS-136

Digital a method of representing information as numbers with discrete values; usually expressed as a sequence of bits.

DOS Disk Operating System, the Microsoft predecessor to Windows.

DRM Digital Rights Management.

EDGE Enhanced Data rates for GSM Evolution; effectively the final stage in the development of the GSM standard, EDGE enables theoretical data speeds of up to 384kbit/s. It provides an alternative upgrade path towards 3G services for operators without access to new spectrum, and enables multimedia transmissions and broadband applications for mobile phones.

ETACS Extended TACS; the extension of TACS by the addition of new frequencies.

FCC Federal Communications Commission; the US regulatory body for telecommunications.

FDMA Frequency Division Multiple Access, a transmission technique where the assigned frequency band for a network is divided into sub-bands which are allocated to subscribers for the duration of their calls.

FM Frequency Modulation.

Gbit/s A data transmission rate of one billion bits per second.

Geostationary A satellite in equatorial orbit above the earth which appears from the surface to be stationary.

GHz A frequency equal to one billion Hertz.

GPRS General Packet Radio Service used by 2.5G systems. It is a technology that sends packets of data across a wireless network at speeds up to 115Kbps. GPRS is designed to work with GSM. GPRS is an essential precursor for 3G as it provides packet-based, rather than circuit-switched connections on mobile networks. The as-needed (rather than dedicated) connections reduce the cost of data services. Based on the GSM standard, it is an incremental step toward EDGE and 3G services.

GPS	Global Positioning System, a location system based on a series of US Department of Defence satellites. Depending on the number of satellites visible to the user, it can provide accuracies down to tens of metres.
GSM	Groupe Special Mobile or Global Systems for Mobile Communications, it is a world standard for digital cellular networks that originated in Europe. It operates at 900MHz, 1800MHz and 1900MHz.
HDML	(Handheld Device Markup Language) a language that formats information for mobile phones or handheld computers in the same way that HTML does for PCs. It is considered to be the forerunner of WML (Wireless Markup Language). Most current HDML browsers are capable of interpreting WML sites.
HMI	Human Machine Interface.
i-mode	Information Mode, a packet based mobile phone service from Japan's NTT DoCoMo. i-mode uses a simplified version of HTML rather than WML and delivers Internet services to subscribers.
IMT-2000	The family of 3G technologies approved by the ITU. There are five members of the family.
Internet	A loose confederation of autonomous databases and networks. Originally developed for academic use the Internet is now a global structure of millions of sites accessible by anyone.
Intranet	A private network which utilises the same techniques as the Internet but is accessible only by authorised users.
IP	Internet Protocol.
IPR	Intellectual Property Rights.
Iridium	A low-earth-orbit satellite communications system developed initially by Motorola.

ISDN	Integrated Services Digital Network.
ISP	Internet Service Provider.
ITU	International Telecommunications Union.
IVR	Interactive Voice Response.
LAN	Local Area Network.
LBS	Location Based Services. Location technology allows a user to determine their whereabouts relative to a particular point of interest or another person. By combining mobile content with location technology, relevant location based services can be made available.
L-Commerce	Location-based e-commerce that responds to a user's physical location.
Line of sight	The connection between communication devices in which there are no obstructions on a direct path between transmitter and receiver.
Mbit/s	Megabit, a data transmission speed of one million bits per second.
MHz	Megahertz, a frequency of one million Hertz.
MMS	Multimedia Messaging Service, an evolution of SMS, MMS goes beyond text messaging offering various kinds of multimedia content including images, audio and video clips.
MNO	Mobile Network Operator.
MP3	A digital audio encoding format used by small digital devices that store and play music.
MVNO	Mobile Virtual Network Operators. A company that buys network capacity from a network operator to offer its own branded mobile phone and value-added services.
Narrowband	A bandwidth of 64kbit/s or less.

NMT Nordic Mobile Telephone system.

P2T Push-to-Talk, a 'walkie talkie' function provided with some types of mobile phone, particularly in the US.

Packet-switching A communication system where voice or data is broken down into packets of a given size that are addressed and transmitted over a network to their destination. Packet-switching is more efficient than circuit-switching because network resources are only used when packets are sent.

PDA Personal Digital Assistant, a small handheld device commonly used as a mobile computer or personal organiser. Many PDAs incorporate small keyboards, while others use touchscreens with handwriting recognition. Some of these devices have Internet capabilities, either through a built-in or add-on modem.

PDC Personal Digital Communications; a digital cellular technology developed and deployed uniquely in Japan. A TDMA technology, PDC is incompatible with any other digital mobile phone standard.

Penetration The percentage of the total population which owns a mobile phone.

Phone book A list of personal names and numbers stored in a mobile phone's internal memory or in the SIM card. These numbers can be called by accessing the appropriate memory and making a single key stroke.

PIN Personal Identifier Number.

Pocket PC (formerly Windows CE) an upgraded version of Windows CE that offers greater stability and a new interface. Features include mobile Internet capabilities, an e-book reader, and handwriting recognition.

Roaming A service that enables you to take your phone to another country and be able to use it.

Router A device that interconnects networks that are either local area or wide area.

SIM Subscriber Identity Module, a smart card containing the telephone number of the subscriber, encoded network identification details, the PIN and other user data such as the phone book. A user's SIM card can be moved from phone to phone as it contains all the key information required to activate the phone.

Smartphone/ Webphone A combination of mobile phone and personal digital assistant that features Internet access, simple text messaging, and data services.

SMS Short Message Service, a text message service which enables users to send short messages (up to one hundred and sixty characters) to other users.

SMSC SMS Centre, the network entity which switches SMS traffic.

SMTP over TCP/IP Simple Mail Transfer Protocol over Transmission Control Protocol/Internet Protocol, a standard for sending email over the Internet.

Streaming An Internet derived expression for the one-way transmission of video and audio content.

Switching Routing traffic by setting up temporary connections between two or more network points.

Symbian A company created by Psion, Nokia, Ericsson and Motorola in 1998 with the aim of developing and standardising an operating system which enable mobile phones from different manufacturers to exchange information.

TACS Total Access Communications System (an AMPS variant deployed in a number of countries, principally the UK).

TD-CDMA Time Division CDMA.

TDM Time Division Multiplexing, a method by which numerous calls are combined for transmission on a single communications channel. Each call is broken up into many segments, each having very short duration.

TDMA Time Division Multiple Access, a technique for multiplexing multiple users onto a single channel on a single carrier by splitting the carrier into time slots and allocating these on an as-needed basis. It divides mobile channels into three time slots, increasing data capacity.

TTS Text-to-Speech.

UMTS Universal Mobile Telecommunications System, the European standard for 3G; now subsumed into the IMT-2000 family as the WCDMA technology.

Voicemail A service offered by network operators where calls received when the mobile is in use, switched off or out of coverage can be diverted to an answering service with a user recorded message.

VoIP Voice over Internet Protocol, a standard for sending phone messages over the Internet.

VPN Virtual Private Network.

WAP Wireless Application Protocol; a de-facto standard for enabling mobile phones to access the Internet and advanced services. Users can access websites and pages, which have been converted by using WML, into slimmed-down versions more suited to display on mobile phones.

WCDMA Wideband CDMA, the technology created from a fusion of proposals to act as the European standard for the ITU IMT-2000 family.

Wideband A bandwidth between 64kbits/s and 2Mbit/s.

WiFi A set of standards for wireless local area networks.

Windows CE a version of Windows designed to run on PDAs or other small devices. CE was renamed Pocket PC with the version 3.0 release.

WML Wireless Markup Language, is an open standard, supported by most mobile phones, and was developed specifically for mobile applications to control the presentation of web pages on mobile phones and PDAs in the same way that HTML does for PCs. WML is based on XML and enables optimum usage of the limited display capabilities of the mobile phone.

WWW World Wide Web.

XML eXtended Markup Language, a standard for creating expandable information formats that allow both the format and the data to be shared. XML is similar to HTML in that both use tags to describe the contents of a page but while HTML only describes how the data should be displayed, XML describes the type of data.

Index

Printed in the United Kingdom
by Lightning Source UK Ltd.
119553UK00001B/54